A History of Energy Flows

This book presents a global and historical perspective of energy flows during the last millennium.

The search for sustainable energy is a key issue dominating today's energy regime. This book details the historical evolution of energy, following the overlapping and slow flowing transitions from one regime to another. In doing so it seeks to provide insight into future energy transitions and the means of utilizing sustainable energy sources to reduce humanity's fossil fuel footprint. The book begins with an examination of the earliest and most basic forms of energy use, namely, that of humans metabolizing food in order to work, with the first transition following the domestication and breeding of horses and other animals. The book also examines energy sources key to development during the industrialization and mechanization, such as wood and coal, as well as more recent sources, such as crude oil and nuclear energy. The book then assesses energy flows that are at the forefront of sustainability, by examining green sources, such as solar, wind power and hydropower. While it is easy to see energy flows in terms of "revolutions," transitions have taken centuries to evolve, and transitions are never fully global, as, for example, wood remains the primary fuel source for cooking in much of the developing world. This book not only demonstrates the longevity of energy transitions but also discusses the possibility for reducing transition times when technological developments provide inexpensive and safe energy sources that can reduce the dependency on fossil fuels.

This book will be of great interest to students and scholars of energy transitions, sustainable energy and environmental and energy history.

Anthony N. Penna is Professor Emeritus of Environmental History at Northeastern University, USA. He is author of numerous titles, including *The Human Footprint: A Global Environmental History* (2015, 2nd ed.), *Natural Disasters in a Global Environment* (2013) and *Nature's Bounty: Historical and Modern Environmental Perspectives* (1999).

A History of Energy Flows

From Human Labor to Renewable Power

Anthony N. Penna

Routledge
Taylor & Francis Group
LONDON AND NEW YORK

earthscan
from Routledge

First published 2020
by Routledge
2 Park Square, Milton Park, Abingdon, Oxon OX14 4RN

and by Routledge
605 Third Avenue, New York, NY 10017

First issued in paperback 2021

Routledge is an imprint of the Taylor & Francis Group, an informa business

British Library Cataloguing-in-Publication Data
A catalogue record for this book is available from the British Library

Library of Congress Cataloging-in-Publication Data
A catalog record for this book has been requested

ISBN 13: 978-0-367-77706-7 (pbk)
ISBN 13: 978-1-138-58829-5 (hbk)

Typeset in Goudy
by Apex CoVantage, LLC

To members of my immediate family
Christina Penna Turner, Matthew David Turner,
Olan Anthony Turner and Gregory Michael Penna

Contents

Figures

Introduction and acknowledgments

This book aims to focus attention on energy flows as a key concept in studying premodern, modern and contemporary history. Formerly the exclusive domain of the natural sciences, the current climate crisis engages scholars across many disciplines to critically analyze energy use in the context of social, economic and political systems. For environmental historians who study climate instability, deforestation, ocean acidification and mass biodiversity extinction and many other unintended consequences of climate change, energy history has become a fast-growing subfield of the discipline. As distinguished ecologist and geographer, Vaclav Smil has noted, historians traditionally ignored the "use of energy as an essential explanatory variable . . . This omission is indefensible if one seeks to understand all of the fundamental factors shaping history."[1]

Since the origins of our planet more than 3 billion years ago to the evolution of *Homo sapiens* in the last seconds of Earth history, all living matter from bacteria to humanity depend on the Sun's energy. It is the primary building block of humanity and a basic foundation of society. Most premodern and all technologically advanced societies depend on energy in the form of light and heat from that star's great big hydrogen fusion reactor. As "children of the sun," we transformed its energy into calories for humans and animals to perform work.

Human labor and animal labor are renewable forms of energy. Both convert food and fodder into calories to do work. Wind, flowing water, sunlight, wood and trees are also renewable. Planting the latter and giving them time to grow make them renewable. Fossil fuels (i.e., coal, oil and gas) and fuels, such as uranium, undergo fission or fusion to release energy. Since these energy sources take millions of years to form from organic material, they are classified as nonrenewable.

A History of Energy Flows: From Human Labor to Renewable Power traces the long and complex history of energy flows from early organic-based energy systems to advanced renewable technologies. Historical analysis uncovered both the positive and negative outcomes of using different sources of energy while emphasizing persistence and change. Urban horses contaminated city streets and walkways with manure and urine daily. Their replacement with gasoline-powered vehicles was heralded as a solution to urban blight. Now, we know that this transition from an organic to a mineral energy source released millions of tons of carbon dioxide, a major contributor to climate change. Although horses, oxen and mules continue as the "beasts of burden" in developing countries, mineral energy sources dominate the industrial world.

Three parts titled the organic energy regime, the mineral regime and the renewable energy regime provide the organizational structure of Chapters 1 through 9. Chapter 10 focuses on battery storage for power generated by renewable wind, sunlight and hydrogen fuel to replace fossil oil in the world's growing transportation sector. Together, this classification highlights the fragmented nature of decisions that define continuity and change in energy systems. As the space held by organic energy regimes (i.e., human and animal labor, wood, and the early uses of wind and water) contracted and a mineral regime (i.e., coal, oil, manufactured and natural gas, and nuclear) expanded, the former did not disappear. In much of the developing world, organic forms of energy persist. Uneven development around the world determined the uses of various energy sources such as human and animal labor, wood, coal, oil, gas, wind and water.

In the developed world, expectations exist among policymakers that a renewable energy regime (i.e., hydropower, wind and solar power) will result in the contraction of mineral energy, reducing carbon emissions and slowing global climate change.

The transition from one energy regime to another was often uneven. The transition from wood to coal was driven by economic conditions and the discovery of a new and seemingly inexhaustible energy source. Others, such as nuclear power, resulted from deliberate and coherent policies, the consequences of which were unintended or unexpected.

Advances in the technology of wind and solar also point to imperative reasons for continuity and change in energy systems. What humanity does in both policy and mitigation to transition away from centuries of fossil fuel dependence will remain one of the great questions facing the future of Earth.

The organic energy regime

The domestication and breeding of animals represented a major energy transition for humans who previously represented the sole biological converter of food into calories. By metabolizing food for work, domesticated animals provided large amounts of energy, releasing many humans from the arduous tasks of daily life. Oxen and horses substantially increased the amount of mechanical energy available to humans, with horses providing 10 times more energy to do work than humans. Animal power lifted water from wells, irrigated the land for cultivation, plowed the fields for planting and turned millstones to grind wheat into flour.

Although linking humans and animals to perform work represented a breakthrough for humans, substituting these much larger biological engines to do work required the cultivation of land to produce fodder in the form of grasses and seeds. They also consumed corn and wheat, food eaten by humans, with higher caloric value. Animal labor required 4,000 to 5,000 calories a day to perform work so that one horse could do the work of five humans. Some societies, however, made progress in many places without the benefits of animal energy for centuries. Yet, a combination of human and animal power created a more energy-intensive environment.

For centuries, wood, a more efficient fuel source, provided heat and light. Wood, cut using the energy gained by the metabolized food of human–animal labor, powered the engines of the early manufacturing-industrial revolution. Milled wood built the factories, the waterwheels that turned a river's energy into power and the vanes that captured the energy of the wind. Burning wood, not coal, powered the early stages of industrialization. For much of human history an organic energy regime prevailed.

Wood became a universal material. Men, women and children sat on wooden chairs and ate off wooden plates with wooden spoons at wooden tables. They slept in wooden beds. They spun wool and flax into yarn and thread on wooden spinning wheels. Carpenters fashioned cut wood into handles and grips for metal tools. Their most important machine tool, the wooden lathe, was made of movable wooden parts. Wooden vats and barrels were used to brew and store liquids. As late as the nineteenth century, the steam engine functioned using many wooden parts, including gears, pumps and boilers. For others, they made carts, wagon wheels, buckets and washtubs. They hollowed out logs and connected them with wooden pegs creating pipes to carry drinking water to emerging cities.

The energy of human labor made the major capital investment profitable. Through exhausting and dangerous labor, men using shovels, hammers, and drills dredged the bottoms of major rivers. Other men with wheelbarrows and assisted of horse-drawn wagons hauled out tons of sediment and rock from river bottoms. Tons of gunpowder blasted away ledges to widen canal systems. Exploding gunpowder was a constant hazard for workers, causing injuries and amputations for some and death to others. After dredging and excavation adding width and depth, stonemasons built retaining walls.

Capturing the power of wind and moving water added value to the organic energy regime. In terms of capital investment, harnessing the power of water required the construction of dams, mostly made of wood, to contain flowing water and create holding ponds and reservoirs as power reserves. The dredging of shallow impediments; the digging and maintaining of canals, locks, spillways; and the construction of floodgates, requiring human, animal and mechanical labor, added to the costs.

Textile mill construction consumed large quantities of lumber for framing, flooring, siding as well as a mill's basic wooden machines of spindles and looms during the early industrial era. Lumber with some iron reinforcement to bolt the machines together represented the primary material for the wheelhouse, the wheel and buckets or paddles and gears. Large quantities of cut timber held riverbanks and canal sidings in place.

Economic factors, including depleted forests and the rising transportation costs for wood cut in distant places, led to a change in policy choices. The discovery of plentiful and cheap coal initiated the emergence of a mineral energy regime and the rise in metallurgy as a replacement for wooden products.

The mineral energy regime

Initially, coal-powered steam engines were used to drain water from mines, work done previously by human- and horse-powered windlasses, a wooden device to lift coal from a shallow mine. Some mines, located near ironworks, replaced wood with cheap coal-to-power steam engines in the smelting of metal. Coal's economic benefits encouraged British manufacturers to bring coal a distance from Newcastle to their ironworks. Steam-powered railcars carried coal to ironworks and finished products from factories to consumers. The superiority of coal-burning steam engines began the long, uneven transition from expensive wood and waterpower.

Coal-burning railway locomotives and steam vessels replaced river flatboats and barges while coaling stations located strategically around the world linked markets of finished goods with suppliers of raw materials. The strategic location of coaling stations for steamships created a form of economic imperialism in Africa and Asia. In the process, local self-sufficient economies collapsed and were replaced by single-crop economies of rubber, sugar, tea, coffee, tin and many others. These economies became victims of a volatile global economy.

This transition differed across space and time. In the United States, virgin forests provided plentiful wood, so not until the end of the nineteenth century did significant change occur. Human, animal, wood, wind and water sources of energy dominated the American landscape until the transformation of water into steam became commonplace in the twentieth century. Combustible materials and boilers and turbines fabricated in ironworks combined to complete this chemical process. In Great Britain and on the continent, forests had been nearly exhausted much earlier. Fossil coal replaced wood by the beginning of the eighteenth century.

Energy transitions may result in abandoning one power source and replacing it with another. A more common practice is a layered one in which a new energy source is added to the mix and over time becomes dominant by releasing more energy to do work. As noted earlier, one horse did the work of five men. Four 52-gallon barrels of oil produced an equivalent amount of electricity as a ton of coal. Burning 714 pounds of coal will power a light bulb for a year, while burning 143 pounds of natural gas will accomplish the same result. Burning 0.035 pounds of natural uranium will do the same.

By 1880 CE, wood, thatch, straw and other biomass gave way to coal as the world's main commercial energy source. Produced in only a few countries including Great Britain and the United States, coal mining increased domestic consumption for heat and light and industrial production for iron and steel. In each decade after 1880, coal mining output increased 50 percent. Burning coal heated boilers that turned water into steam that turned turbines to heat households, power factory machinery and power ships on the high seas. Initially used to drain water from deep mines, coal-burning steam engines powered the battering rams that released "liquid gold" and natural gas from their deep subterranean cavities. With the discovery of oil, the remaining remnants of the organic energy regime receded further into the background of industrial economies.

The visibility of organic energy, however, remained viable during the transition. Men labored by collecting percolating oil from the pits they dug to contain its spread over the land. Wooden vats contained crude oil; wooden barrels loaded on wagons, pulled by horses and led by teamsters, were transported to barges on rivers and railcars on roads. Humans, horses and wood converged to accelerate the transition to a petroleum-based economy compatible with the use of fossil coal. As with each transition, technical advances accelerated the development and utility of petroleum use. Welded pipe eliminated the need for humans, horses and wooden containers. By eliminating the horse-drawn wagon for transportation, pipelines, pressurized by steam-powered pumps, diminished the power of teamsters

Coal-fired factories produced plate steel for automobiles, trucks, ships, submarines, and airplanes. Coal also became instrumental in the extraction and refining of crude oil. In the early decades of the twentieth century, from 1900 to 1940, the consumption of oil grew by 50 percent with the United States, producing two-thirds of the world's production of 1 million barrels annually. By tapping into global supplies in Southwest Asia, Saudi Arabia, Iran, Iraq and the Arab Emirates, Britain, France and the United States used its political and economic might to capture control of foreign oil supplies. The effects proved to be transformative as refined petroleum provided the liquid fossil fuel to power modern mobility.

The introduction of petroleum into the energy mix changed the relative dominance of coal. Coal maintained its primacy in manufacturing iron and steel by using its vast reserves to produce coking coal produced in beehive coke ovens. At the same time, electrical utilities consumed large reserves of coal for heat and light. Burning coal and wood in tightly controlled beehive ovens produced coke, this new energy-intensive substance for the iron and steel industry, but wasted its chemical by-products. Scientific knowledge and technological applications in the last decades of the eighteenth century improved the chemical process of distillation by capturing coke's chemical by-products in a retort. Industrial distillation is a separation technology designed to make a more efficient use of energy sources. Burning coking coal in a retort and capturing its gas made fostering a more efficient-energy regime possible. The creation of the manufactured gas industry established direct links to coal mining, coking coal and its by-product, coal gas.

Once again, rather than a grand narrative to describe energy transitions, the manufactured gas industry came about as a reflection of socio-technical conditions. The social context included the need to provide lighting for city streets. The technical context included the rapid development of chemical distillation and the expansion of manufactured gas usage in household appliances once electric lights replaced its use in street lighting. Competition from a new generation of electrical appliances and the rapid growth of the kerosene industry led to the demise of the coal-gas industry in the following decades.

Initially, wildcatters and petroleum industry executives viewed natural gas as a waste product in the drilling for crude oil. Making the transition from manufactured gas to natural gas faced barriers related to investments in mines, buildings and infrastructure. Coal reserves, coal-gas plants, pipelines and local distribution centers, as well as the production and installation of gas appliances for households and commercial enterprises, added to the sunk costs of investors and owners. In addition, pipeline technology posed as a barrier to distribution. Gas explosions caused injuries, death and property damage. As the demand for natural gas grew, pipeline technology improved rapidly. Oxygen-acetylene welding to the joints of pipes, first used in 1911, prevented gas from leaking under high pressure and minimized explosions. The invention of seamless electrically welded pipe ensured a smooth pipe inside, accelerating the movement of natural gas from producer to distributor to customer. Technological breakthroughs signified the beginning of the end for manufactured gas.

Nuclear power belongs in the mineral energy regime. Nuclear power requires mining the world's nonrenewable supply of uranium, the only fissionable element from the periodic table that contained the neutrons capable of a chain reaction. Although Earth contains millions of tons of uranium, its supply, like that of coal, petroleum and natural gas, is a depleting energy source. Currently, some reactors are running on mixed oxide blends of uranium and plutonium. The next generation of plutonium breeder reactors to fuel future reactors, and the third generation of thorium reactors may solve the depletion problem. Those developments may be decades away. In addition, the full life cycle of building a nuclear power plant requires cement and industrial-grade steel. Both emit tons of carbon dioxide in the mining and manufacturing process.

Despite the promise of nuclear fission to produce boundless amounts of energy, much cleaner than burning coal, research and development focused on developing a nuclear bomb before Nazi Germany achieved a similar goal. Military objectives, rather than peaceful uses of nuclear technology, dominated world affairs in the World War II era. Once the West's former ally, the Soviet Union, detonated a nuclear bomb, the threat of a global war focused the attention of world leaders. A Cold War atmosphere resulted and a nuclear arms race between the world's two superpowers, the United States and the Soviet Union.

While the nuclear arms race dominated world affairs and an increasing number of nations acquired and processed fissionable material to manufacture nuclear bombs, many of these nations also invested in nuclear power plants to produce electricity. A goal among civilian advocates of nuclear power was that electricity would become available to households so cheaply that it would be pointless to meter. Today, the global nuclear power industry operates 450 reactors. In the West, a pause in nuclear power installations reflects reservations over safety. Major catastrophes at Chernobyl, Ukraine, in 1986; Three Mile Island, United States, in 1979; and Fukushima, Japan, in 2011 were believed by the public at large to spread atmospheric radioactive material across national boundaries.

However, comparing the mortality rates per trillion kilowatt-hours of energy produced by coal far exceeds deaths caused by nuclear energy. Burning coal in China results in 170,000 deaths per trillion kilowatt-hours. Nuclear power caused 90 indirect and direct deaths from radiation exposure and increased cancer risks.[2] China produces 75 percent of its electrical energy from coal combustion.

Its construction of 40 percent of the world's nuclear reactors (26 out of 65) reflects its increasing demand for electricity and its efforts to curb its dependence on fossil fuels and reduce the country's air pollution. Given China's current capacity, it eliminates an estimated 67 million tons of carbon dioxide that would have entered the atmosphere by burning fossil fuels. India's growing population and the modernization of its economy will require greater amounts of electricity. Currently, it lags behind China in nuclear power planning. Stymied by its modest reserves of uranium, India produces 4 percent of its energy from nuclear sources while the per country average is approximately 16 percent.[3]

The renewable energy regime

The kinetic energy of water reduced dependence on human and animal labor during the ancient world and the various processes of modernization during the era of manufacturing. The power of waterwheels transformed the production process by replacing human and animal work with machines. The circular motion of flowing water turned water wheels to activate the many machines that used circular motion to transform raw cotton into cloth. The modern world of hydropower commenced with the building large dams to stop and store the flowing water of large rivers. Within the bowels of these hydroelectric dams, powerful turbines awaited the release of millions of gallons of water to produce electrical energy. They delivered this energy across a complex system of grids to power machinery great distances from its source and to illuminate cities, towns and households.

During the twentieth century, large rivers and their tributaries with rapidly flowing water became synonymous with renewable energy. Melting winter snowpacks, some originating in the highlands and mountain ranges, fed rivers seasonally. They became ideal locations for hydroelectric dams. During that century, governments and utilities across the world built as many as 45,000 hydroelectric dams promoting its renewable benefits. It controlled seasonal flooding, managed irrigation and improved river transportation. However, the world's largest rivers became enormous reservoirs of still water, raising temperatures and reducing the flow of silt that renewed the land's ecological balance. The displacement of populations dependent on a river's aquatic abundance and transportation system mounted steadily with each new construction. In all, at least 30 million people lost their homes and livelihoods. Other estimates placed the number of displacements much higher at 60 million people. Brazil, China, Canada and the United States became the world's largest producers of hydroelectric power and the greatest number of resettlements. Although the century of dam construction in the United States has ended with the decommissioning of more than 300 large dams, China and Brazil continue building large dams to modernize their economies.

Regarding the potential power of solar energy, the power of the Sun dwarfs all other sources of renewable energy. This vast hydrogen fusion reactor warms us every day of every year. It has been doing the same for the last 4 billion years. Each year, we receive 885 million terawatt-hours of solar energy. Each day, Earth gets 165,000 terawatts (1 terawatt-hour equals 1 million megawatt-hours) of energy from our galaxy's Sun. With a global population predicted to reach 10 billion by 2050, harnessing 60 terawatt-hours will provide each person with a few kilowatt-hours (1 kilowatt-hour is 1,000 watts) of electricity each day. Covering a landmass equivalent to the size of Texas with state-of-the-art solar panels in 2050 will provide that few kilowatt-hours each day to the entire world's population. Pointing in that direction, solar power became the world's leader in new electricity generation in 2017, installing 73 gigawatt-hours of new solar photovoltaic capacity. (One gigawatt-hour is 1 billion watts.)

Unlike energy sources that change with fluctuating needs, with availability, costs and the population density to produce power efficiently, the power of the wind has remained a constant throughout Earth's history. Its force sweeps across the land, rearranging the topography, creating dunes in some places and eroding hilltops and mountains in others. Dust storms became a signature event by moving topsoil across continents. Its power and presence were available to humans across time and space. Its energy filled a ship's sails and propelled it forward and the blades of early windmills raised groundwater to the surface. Over time, the increasing size and efficiency of windmills assumed many more functions. The invention of turbines in the nineteenth century set the stage for the modern technology of wind power. Similar to sunlight that diminished and ended each day, the variability of the wind's velocity inspired the efforts to store solar and wind power for timely use.

Storage technology remains at the forefront of future energy reserves. To power vehicles and to generate electricity for households, commercial and industrial facilities hydrogen fuel cell technology, and innovations in battery power may pave the way to a more carbon-free world. Relative to other ways to generate electricity, fuel cells have the advantages of low costs with high efficiencies. As the most abundant element found on Earth's surface, hydrogen, is locked up in water and plants. Despite its abundance, hydrogen is always combined with other elements, such as water. By passing an electrical current through water, it can be separated into its components of hydrogen and oxygen. This separation process is known as electrolysis. A hydrogen-oxygen fuel cell produces electricity, heat and water. Fuel cells will continue to produce electricity and never lose their charge as long as the fuel source is hydrogen. Since hydrogen fuel cells produce no heat-trapping gases, they play no role in warming the global climate system.

Energy and the power that they generate determine the value of batteries. Energy defines the amount of work that can be completed and power tells us how quickly that work gets done. In addition to hydrogen-oxygen fuel cell technology, research and development on long-lasting batteries, some larger than ship containers, offer another opportunity to break our fossil fuel dependence.

Storage batteries eliminate the variability that occurs each day from wind and sun energy sources and regulates the demand for power. Large storage batteries located at central data centers that controlled the flow of power would eliminate the need for utilities to build additional coal-fired and natural gas plants to match the increased demand for power. Unlike conventional power plants that often take time to meet spikes in demand energy stored from wind and sunlight in large batteries would dispatch power immediately and provide another opportunity to break our dependence on fossil fuels.

Conclusion

Two alarming reports in 2018 point out the folly to humankind by continuing to contaminate the atmosphere in carbon emissions. Carbon dioxide rates continue to rise as nuclear power plants close and are replaced with natural gas and coal. Renewable wind and solar sources are growing rapidly but not fast enough to reach zero emissions by 2030. New solar cost US$50/megawatt-hour, more than half the cost of coal, but again scaling up to replace fossil fuels seems unachievable in the time frame recommended by the Paris Agreement of 2015.

First, *The New York Times* devoted its entire Sunday magazine on August 5, 2018, to the topic of climate change. Titled "Losing Earth," by Nathaniel Rich, it noted that the rise of 1 degree Celsius (1.80 °F) since the beginning of the Industrial Revolution in the nineteenth century placed Earth on its current path of rising temperatures caused by our fossil fuel dependency. Rich suggested that the world's climate was positioned to rise another degree Celsius, given present carbon emissions. Approaching a climate system with a global warming of 2 degrees Celsius would, according to climate scientist James Hansen, be "a prescription for long term disaster."[4] Global sea levels would rise several feet and the world's tropical reefs would disappear forever. Further warming of the global climate system by an additional degree Celsius, a realistic minimum, would result in the loss of most coastal cities and the migration of many millions of their residents.

The United Nations Intergovernmental Panel on Climate Change released the second alarming report on October 6, 2018, on the impact of global warming above preindustrial levels. The panel noted that human activities have raised global temperatures by 1 degree Celsius since the Industrial Revolution, with a likely range of 0.8 to 1.2 degrees Celsius. With emissions on their current path, global warming is likely to reach 1.5 degrees Celsius between 2030 and 2052. Human-induced warming is rising 0.2 degrees Celsius each decade from past and ongoing emissions. As the report noted, warming will continue for centuries to millennia and will cause further long-term changes in the climate system, including sea-level rise. If global warming exceeds 1.5 degrees Celsius and above 2 degrees Celsius, some impacts may be long-lasting or irreversible. Sea-level rise will continue beyond 2100 even if global warming is limited to 1.5 degrees Celsius in the twenty-first century; eroding ice sheets in Antarctica and/or the loss of the Greenland ice sheet could result in a multifeet rise in sea level over hundreds to thousands of years, causing the displacement of millions of people. Millions more could become exposed to record heat. Changing our entire energy system may seem impossible, but the history of energy transitions offers reasons for optimism. Major technological transformations can take place over a quarter century. If fossil fuel emissions are phased out by mid-century and replaced by the renewable forms of energy described in this volume, our optimism for a sustainable energy future will become a reality.

Acknowledgments

Writing a chapter on energy history for an earlier book, *The Human Footprint: A Global Environmental History*, encouraged me to write this book. Reaching out to scholars in the field of energy history made me an enthusiastic participant in the ongoing discussion about this newly emerging field. Seminars organized by Professor Ian Miller at Harvard University in 2015–2016 brought together scholars whose current work in energy history provided insights into the role of energy resources as significant contributors to the shaping of modern society. They included Paul Warde, John McNeill, Christoper Jones, Conevery Valencius and Edward Melillo. During the writing, my former colleague at Carnegie-Mellon University and longtime friend, Joel A. Tarr provided source material and suggestions for improving the text. Harvey Green, my colleague at Northeastern University, assisted me in writing the proposal and provided commentary on the manuscript's introduction. Patrick Larkey, a former colleague at Carnegie-Mellon University, provided me with sources related to storage battery technology. My companion, Deborah Addis, a professional writer, played an important role in revising the introduction and, as a skilled photographer, helped in selecting and securing photographs and images for the manuscript. Colin Sargent, my former graduate student in world history, provided preliminary photographs for the volume. James Parker, a current PhD candidate in world history at Northeastern University, created the reference files for the manuscript. Special thanks to Heather Streets-Salter, chair of the department of history at Northeastern, for providing me with an office to think, do research and write during my years as an emeritus faculty member. Henry Giroux, my former student and long-term friend, introduced me to Dean Birkenkamp, editor of Sociology at Routledge who forwarded my book proposal to Annabelle Harris, editor of Environment and Sustainability.

Notes

1 Vaclav Smil, *Energy and Civilization: A History* (Cambridge, MA: The MIT Press, 2017), 430.
2 Nancy Langston, "Closing Nuclear Plants Will Increase Climate Risks," (Active History.ca) January 30, 2019.
3 Zeng Ming, Liu Yangxin, and Ouyang Shaojie, et al., "Nuclear Energy in the Post-Fukushima Era: Research on the Developments of the Chinese and Worldwide Nuclear Power Industries," *Renewable and Sustainable Energy Reviews* 58 (May 2016): 147–156.
4 *The New York Times*, Sunday Magazine on August 5, 2018.

Part I

The organic energy regime

1 Biological converters of energy

Food, fodder and firewood

Introduction

At the end of the world's last Ice Age in approximately 11,600 BCE, global populations did not exceed 10 million people. As the most recent current warming stage in human history, scientists and historians refer to this epoch as the Holocene. Prior to this warming period, the sequence of human evolution, however unclear, proceeded from *Australopithecus africanus* to *Homo habilis*. Further evolutionary history followed with *Homo erectus*, appearing about 1.7 million years ago. This hominid species left Africa in successive waves through Southwest Asia, populating significant portions of the Eurasian landmass. Each of these hominid species either disappeared or continued their evolutionary cycle. *Homo sapiens*, our ancestors, evolved from *Homo erectus* and began its migration out of Africa approximately 150,000 to 200,000 years ago. All previous species of the genetic type, *Homo*, lived a nomadic life as hunter-gatherers, moving constantly in search of eatable plants and animals. Uncooked food sustained these archaic humans and remained their only source of energy for millennia.

Humans metabolizing food into energy in order to work dominated the production of energy for most of human history before the domestication and breeding of horses and other animals. This change was gradual and inconsistent in the world. Many people continued to cut down trees for fuel, hew logs for construction and seed fallow ground without the benefit of animal energy for centuries after domestication – and eventually industrialization and mechanization – were common in parts of Europe and Asia.

The transition from human and animal labor continued for centuries, even as a more efficient fuel source, wood, provided opportunities for the invention of machines that could dwarf the human–animal potential for energy creation. While people today tend to see energy transformation and creation in terms of "revolutions," this energy transition took centuries. In spite of the commonly held assumptions about energy and the Industrial Revolution, the invention of machines for manufacturing and industry were not run on wood's eventual replacement, coal. Wood-powered engines built the factories and mills that housed machines and wheels that harnessed the kinetic energy of rivers and the vanes that caught the energy of winds.

The discovery of fire

The discovery and control of fire by *Homo erectus*, dating back at least 500,000 years represented the first great ecological transition for humans, influencing their evolutionary trajectory. "The discovery of fire and the exploitation of draft animals marked two main changes in the history of technology."[1] Although scant evidence exists about how this discovery and use were made, watching naturally occurring fires caused by volcanic eruptions and falling rocks may provide some of the circumstantial evidence.

One hypothesis developed by interpreting the fossilized remains of large predators in a cave in Sterkfontein, South Africa, suggests another way in which humans gained control over their environment. For many generations, big cats, possibly saber-tooth tigers, much larger than today's Indian tigers, feasted on *Australopithecus africanus*, who had not mastered control of fire. Then, *Homo erectus* arrived on the scene, reversing the age-old tradition of predator and prey.

Another hypothesis put forth by researchers suggests that by possessing the ability to hold a stick and by poking through a fire looking for food, *Homo erectus* learned that sticks in a fire promoted burning. So, adding more sticks kept the fire burning, and carrying a fire stick allowed them to set fires on their own. Keeping and maintaining a fire required planning, the help of others and the expenditure of energy to collect sticks and other combustible biomass sometimes by walking a fair distance. Thinking and cooperation became essential ingredients for success.

The use of fire provided early humans with the ability to cook their food. Cooking made both animal parts and vegetation more digestible and nourishing. Although we do not know how this transition to cooking occurred, recent evidence suggests that cooking meat led to increases in brain size and human intelligence. With climate change and rising temperatures, early cultivators used fire as a readily available *energy support* to light the darkness. It deterred predators, burned away destructive vermin and cleared the land. By engaging in these activities, they unknowingly fused nitrogen to the soil and improved its productivity. Many *energy supports* followed including the domestication of animals and their use in farming and for transport. The invention of the plough, the wheel, the cart and much else accelerated the transition from human to animal energy for work. Population growth became one of its unintended consequences.

The spread of "fire-stick farming" gave humans a readily available tool for clearing the land for cultivation, killing off vermin that spread disease, and hunting. It may also have been used to deter competing humans from invading one's food and shelter. Although the use of the fire stick improved their technical capabilities and health, physical stature and social structure, it increased their ability to convert cooked food into work. Despite this transformation, for a little less than 90 percent of human history, food was the only source of energy for humans, with fire representing 9 percent and working animals, 1 percent.[2] The remainder was animals, water and windmills. All these percentages would change with the transition from gathering and hunting to cultivation and early agriculture encouraged by the warming climate of the Holocene.

In the meantime, the early land practices of *Homo sapiens*

> had profound impacts not only on the fire regimes but also on the landscape vegetation pattern and biodiversity. Commonly, woody, closed-canopy shrublands and woodlands were opened up or entirely displaced by fast-growing annual species. They provided greater seed resources and planting opportunities. These changes also had cascading effects on ecosystem function. In fact, fire-stick farming was probably necessary not only to open up closed-canopy woodlands to create habitable environments but also to reduce catastrophic fires that would pose a risk to humans.[3]

Along with thought processes and social interaction associated with getting and maintaining control over fire, cooking around an open fire unleashed higher food energy. It provided *Homo erectus* with the energy to travel farther, out of Africa into Southwest Asia, with a greater capacity for work and a fitness advantage over other archaic humans. Stored cooked food, a new feature, suggested a plan for distribution within a group. Gathering around fires for eating and socializing, as well as using fire to fend off predators, may have encouraged group cohesiveness.

Some scholars suggest that cooking with fire may have stimulated the development of bigger brains and bodies, smaller teeth and leaner limbs that prompted easier mobility and the social development mentioned earlier. Cooking may have extended the lives of these early humans. Using fire extended their home range into colder regions of Eurasia and allowed them to use the energy of fires to alter their surrounding environments. New social organizations, including the suggestion of a "grandmother" hypothesis regarding the nurturing of children, released parents for other productive activities.[4]

For archaic humans, the primary energy source derived from their labor of gathering and scavenging and then transitioning to overhunting large megafauna and eventually smaller mammals. They consumed nuts, fruits, grains, edible vegetables, animal flesh and fish, converting the food into the energy for work. Food is to humans and animals what gasoline is to mechanical engines. The major difference between humans and machines, however, requires humans to expend energy in search of the food that they need to survive. Archaic humans spent the majority of their waking hours securing the food that they needed to survive. If they failed in this quest because of injury, poor judgment, an inferior skill set or environmental conditions, they died.

Over time, successful technical innovations, including throwing projectiles, such as clubs, sharpened spears, bows and arrows, and strategies for corralling prey for the kill increased efficiency. At the same time, overhunting and burning off the vegetation to make the hunt more successful could alter the environment in fundamental ways. Regardless,

> their food base remained very limited and populations were therefore extremely sparse. As with animal species, moreover, there were often major

> seasonal fluctuations in the quantity and type of food available for consumption, and the size of the population which could be sustained was further limited.[5]

Throughout much of *Homo sapiens*' early history, Earth's climate 120,000 years ago was warm, suggesting an abundance of food and fiber. As the climate cooled in anticipation of the last great Ice Age that lasted for about 90,000 years with intermittent warming phases, *Homo sapiens* migrated out of Africa into the more temperate region of southwest Asia. By 50,000 years ago, they reached New Guinea and 40,000 ago settled in Australia. Extending their nomadic range to Siberia, they crossed the ice-covered Bering Straits 20,000 years ago and began their movement southward into the Americas.

Throughout this long evolutionary history, the human population remained relatively small, numbering no more than 1 million persons 100,000 years ago. Despite its size, however, megafauna collapse followed the encroachment of humans. As geographer Jared Diamond has pointed out, the North American Plains 15,000 years ago looked like the African Serengeti Plain today, with "herds of elephants and horses pursued by lions and cheetahs, and joined by members of such exotic species as camels and giant ground sloths."[6]

The invention of agriculture

The invention of agriculture in multiple locations around the world during the warming of the Holocene resulted in an increase in the global population. At its beginning, in 11,600 BCE, approximately 4 million people inhabited the planet. Between that date and 5000 BCE, another 1 million people inhabited the planet. Population growth accelerated from the 5 million noted earlier to 15 million in 3000 BCE and to between 170 and 250 million in 1 CE. Later, humans would make the transition to a new energy regime. As water and windmills became available in limited quantities, they represented the only nonbiological sources of mechanical energy.[7]

With the onset of the Holocene 11,600 BP, humans experienced a changing climate with rising temperatures. With global carbon dioxide reaching 280 parts per million, an important milestone in plant growth, humans began to cultivate edible plants. Through a protracted and uneven sequence of experiences, cultivation replaced gathering. The age of agriculture encouraged settlement; the development of a common language, rituals and rules; and the beginnings of social hierarchies.

Despite the approximate 200,000-year history of *Homo sapiens*, global warming and the human invention of agriculture became prerequisites for the development of civilizations. Human settlement, rather than lives on the move in search of food, led to the development of institutions over long historical periods that possessed many of the characteristics that we identify today as political institutions, systems of monetary and economic exchange and social hierarchies and norms.

Human energy from food

Throughout this long history, humans converted caloric energy into work, inefficiently for most of its history, despite improvements in their physical and technical capabilities. What we do know is that a person receiving 3,500 calories of food each day can work longer and harder than a person receiving 2,500 calories. If a lower caloric diet persists and extends to the entire community, then big differences will separate one community from the next in terms of its ability to perform useful work:

> A diet providing 2,500 calories a day represents a food intake 71 percent as large as a diet consisting of 3,500 calories. But if food intake among a working population were permanently reduced from 3,500 to 2,500 calories on average, the ability to carry out useful work would fall not by 29 percent but by 50 percent because the first 1500 calories of a daily diet are needed to meet basic metabolic needs. The quantum of energy available for work is therefore halved from 2,000 to 1,000 if daily calorie intake falls from 3,500 to 2,500 calories.[8]

One thousand five hundred calories a day per person may sustain life but not a working life.

If humans need about 75 percent of their energy from food intake, just to stay alive, only 25 percent is available for work. No essential change in this ratio occurred until the Neolithic Age (8000–5500 BCE) when the domestication of animals created a higher degree of energy efficiency. With the availability of animal muscular strength, humans were free to engage in activities other than searching, gathering and consuming food. From that time onward, harnessing other sources of available energy liberated humans from the inefficiencies of converting food into muscular strength:

> There have been two periods of particularly major change: the centuries around the birth of Christ, when water power was harnessed, and the last three centuries, when steam power, internal combustion engines, and nuclear power were exploited and new scientific advances made it possible to use earlier sources of energy more efficiently.[9]

As humans made the transition from cultivation to agriculture, eating enough nutritional food determined one's work output. Variations from one community to another and one person to another would suggest substantial differences in living standards. For example, in antiquity, with high population densities, Roman and Greek diets varied across the socioeconomic spectrum. Urban high-income groups consumed pork meat, while the poorer ancient population ate vegetarian food primarily without a significant difference in calories.[10]

Energy from animals: oxen and horses

Mules provide five times more energy than a human, one ox 7 and a horse 10 times as much energy. In Chapter 2, we will see that watermills made all these sources insignificant by comparison. Energy from animals represented the first major transition for humans. For most of their history reaching back 4 million years, humans were the sole biological converter of food for work. Animals substantially increased the amount of mechanical energy available to humans. Animal power was used to lift water from wells to irrigate land formerly unsuitable for cultivation and for grinding grain into flour. The use of the oxen and donkey for plowing began 4000 to 3000 BCE and that of the horse possibly 1,000 years later. The invention of the wheel about 3000 BCE created a system of transport unknown beyond southwest Asia and Europe. As will be noted later, primarily governments and elite members of society owned warhorses in the ancient world. Oxen were working animals in European antiquity to 1500 BCE.

Estimates suggest that humans tamed horses for leather and manure at the beginning of the Holocene approximately 11,600 BP and before they were domesticated and used for work and transportation. Domestication, meaning controlling their reproduction and making horses into power sources took millennia. As one of the few animals with herding and following instincts, horses could be trained using control mechanisms, such as the metal bit or some other

Figure 1.1 Oxen walk up a slanting plate to turn a wooden gear system beneath them, sending the rotary movement to the milestone inside the building for crushing cornmeal; Upper Italy, 1600 CE.

resilient material placed in the gap between its teeth and connected to a bridle that controls their direction and speed. Recent studies by Russian and Ukrainian archaeologists point out that bit wear from the fossilized teeth of horses found in Dereivka, Ukraine, suggests that people began riding horses by 4000 BCE.[11] Humans millennia ago probably learned that horses "tended to bond with individual riders, since horses can recognize individual voices and intonations. Food, even sweets, might have been a reward."[12]

It cannot be emphasized more about the impact of horses on human society. In the agricultural world, horses replaced human labor. Before their domestication, people traveled on foot carrying their belongings regardless of the weight. Such traveling required frequent stops and took many hours and days to reach one's destination. Riding on horseback or in a cart, wagon or chariot pulled by one or more horses altered people's concept of distance. Speed changed their concept of time as well, promoting increased interactions between rural and urban life as well as extending regional trade among previously isolated cultures. "Horseback riding and horse-drawn chariots may have also played a role in the initial spread of the Indo-European languages, a language family that ultimately gave birth to English, French, Russian, Hindi, Persian, and many other tongues."[13]

Horsepower in antiquity

In ancient Roman society, given the large number of slaves, estimated to be between 30 and 35 percent of the population, the domesticated horse did not replace human labor. When necessary, Roman citizens used oxen for plowing. In Republican and Imperial Rome, horses excelled in their ceremonial role as *equus publicus*, for riding and warfare. Stud farms owned by the nobility and managed for by grooms and handlers, some of whom were slaves, bred and trained thoroughbreds for chariot racing and as cavalry mounts and imperial horse guards. In battle, chariots carried a driver and warrior, who stepped off to fight on foot once the battle commenced. The chariot drivers then retreated a safe distance to watch and care for their steeds. At the huge Circus Maximus in Rome, with a seating capacity of 260,000, spectators marveled at the four-horse chariot races.[14]

However, horse-pulled chariots preceded Roman society by a millennium or more. Written and pictorial evidence placed horse-drawn chariots in Anatolia (modern Turkey) as early as 1900 BCE. According to the Greek historian Herodotus, the Scythians, a Eurasian nomadic culture, waged war on horseback. His description of their culture, social structure and military prowess in 450 BCE follows:

> A people without fortified town, as the Scythians do, in wagons which they take wherever they go, accustomed one and all, to fight on horseback with bows and arrows, and dependent for their food not upon agriculture but upon their cattle: how can such a people fail to defeat the attempt of an invader not only to subdue them, but even to make contact with them?[15]

In ancient Chinese society, horses may have been domesticated about 1500 BCE and used in chariots but not ridden until 300 BCE. China's nomadic neighbors, the Mongols, adopted the cavalry horse and threatened China's northern and western borders. This horse was bred first in ancient Persia. Throughout its ancient history, Chinese emperors waged a systematic struggle to protect itself from Mongol invaders by buying many thousands of horses from them and others. These transactions represented an important exchange of commodities across Eurasia.

The Chinese emperors sold silks and then tea to their neighbors and potential adversaries for horses bred by people who spoke different languages, inhabited different spaces, developed and maintained separate cultural and social institutions yet remained linked together by trade. Why did the Chinese not domesticate and breed their own horses? Scholars suggest that land set aside for fodder would have limited the space needed to grow rice for its growing population. Second, the planning and maintaining of rice paddies precluded the need for horses. So, a horse culture that developed in other parts of the world failed to develop in Imperial China. Finally, Chinggis Khaan (1162–1227 CE), with an army of 30,000 in which each warrior possessed two horses, breached the Great Wall and conquered China in 1279 CE.

As China's famous scholar, Sung Ch'I (998–1061 CE) noted:

> The reason why our enemies to the north and west are able to withstand China is precisely because they have many horses and their men are adept at riding; this is their strength. China has few horses, and its men are not accustomed to riding; this is China's weakness. Those who propose remedies for this situation merely wish to increase our armed forces in order to overwhelm the enemy. They do not realize that, with horses, we can never create an effective military force[16]

In Europe, after the collapse of the Roman Empire in 476 CE, the use of stirrups allowed mounted warriors to throw a lance and shoot arrows with a bow. Legendary acts of chivalry by heavily armored feudal knights became commonplace. Such armor required the breeding of stronger and taller horses. Feudal lords and the knights in their employ ruled vast land areas inhabited by serfs and their tenants. Fighting on horseback represented one of the favored strategies by the descendants of the invading barbarians who had defeated Rome. By 1000 CE, armed horsemen in the employ of wealthy landowners ruled much of Europe. Near the end of that century, seeking to extend their dominion in the east, they initiated what became the Crusades, the first of which began in 1096 CE and ended with the capture of Jerusalem on July 15, 1099. The Crusaders held the city for almost a century until 1187 CE, when Muslim powers took it back. Successive crusades followed.

Horsepower in the medieval and modern world

As is clear from these descriptions of the horse in antiquity and later, its role and function covered a range of activities, from ceremony to sport and most commonly as a warhorse. It would continue to play this role through the end Great War (1914–1918) into World War II (1939–1945). Estimates suggest that all the combatant nations in the Great War deployed a total of 16 million horses with 8 million deaths by the war's end. Germany used 2.7 million horses in World War II, with 1.8 million succumbing in battle. Then, its wartime activities would diminish and end with the total mechanization of war in the mid-twentieth century.[17]

None of these historical developments in the use of horses led to a reduction in human labor on the farm. The use of oxen to plow a field in preparation for planting began in at least 4000 BCE, but the increased use of heavy horses in medieval Europe became more common with inventions that permitted horses to work more efficiently. Europeans adopted the heavy plow in the 600s CE and invented the nailed horseshoe in about 800 CE that increased their use across Europe for transport and carrying loads. Adopting the collar harness and breast strap in the 900s CE allowed horses to pull three to four times their weight. However, the process of replacing oxen with horses was a slow one since horses were viewed as best suited for riding and military uses and not for mundane tasks such as plowing. Also, horses consumed oats, a more expensive option than grass, for energy. Once farmers introduced the three-field system of rotation, oats became a less costly alternative. Twenty years after the Norman invasion of Britain in 1066 CE, horses provided one-fifth of its power supply. From 1000 CE to 1300 CE, horse populations grew to equal oxen as power suppliers. By 1600 CE, horses became the main source of animal power until steam engines replaced them as the primary energy source in mid-1900 CE.[18]

However, replacing human strength with animal power came with costs as well as the benefits noted earlier. Horses that increasingly replaced oxen needed about three to five acres of meadow for fodder that they metabolized into energy. The same was true for oxen. Fertile land, used in this way, eliminated its use for crops eaten by humans. Aside from providing large amounts of mechanical energy, animals supplied manure to fallow land.

> To have a good manured land the ratio was more or less a big size animal per hectare . . . this [kind] of agricultural needs a capital, an animal. The word *capital* does derive just from the Latin word to single out an animal, a *caput*.[19]

Although the working animals from this earlier preindustrial age were much smaller than their successors, who, through selective breeding, were much larger, these biological engines needed about 4,000 to 5,000 calories a day from fodder to perform work. And their ratio to humans was one animal for every five humans.

On average, mature horses over time would replace oxen in the fields were about 10 times stronger than men. They, along with oxen, changed the energy equation to such an extent that draught animals hitched to plow and working in the field reduced human labor by one-half and in some tasks by two-thirds. Now domesticated animals complemented human labor and would begin to fulfill the same function.

We do not possess accurate figures about the amount of work domesticated oxen and horses performed until the Middle Ages. Data from the ancient Greco-Roman world are largely missing. For England, evidence from the Domesday Survey (1086 CE) suggested "the potential supply of power available from draught animals was five times that available from humans in 1086."[20] For much of the work, human strength continued to dig, hoe, mow and cut grain and corn stocks for at least eight more centuries. By 1300 CE, however, animals supplied four times as much draught power as humans and animal numbers grew accordingly from 1086 to 1300 CE, from about 162,500 horses to an estimated 400,000 and for oxen from 650,000 to about 800,000. And the animal numbers continued to grow in successive centuries, approximating three to four times greater than England's 13,500 mills 9,000 were water and 4,500 were windmills. Earlier, at the time of the Domesday Survey in 1086 CE, owners operated 6,100 small watermills.[21]

Specific energy-consuming activities by animals provide evidence of their viability and importance in productive agriculture. Again, the evidence from the ancient world is largely missing, so we are dependent on data from late 1000 CE medieval England. There, before the widespread introduction of horses for plowing, a typical plow team consisted of eight oxen and two men, one man holding the plow and a second man driving, using a whip to move the team of oxen forward:

> Assuming a 0.52 horsepower per ox, then the power potential represented by the eight oxen was 4.2 hp, compared, to, again the 0.16 hp represented by the two males, in this case slightly better than a 25 to 1 return on the original human energy investment.[22]

If we add to this endeavor, the minimal cost of pasturing and caring for the oxen as well as building and repairing the plow, a ratio of twenty to one of animal power over manpower seems to be a reasonable calculation.

As the data suggest, this eleventh-century world in England and probably throughout Europe was a mixed one in which most families owned at least one ox and with some owning a horse. By early 1300 CE, horses became more available as work animals. Although horses cost more than oxen, their versatility outweighed the initial costs. Unlike oxen that were available for plowing, horses performed this function as well as hauling, riding, and carrying packs. Also, there was the variable of speed, where horses plowed faster and hauled goods at about twice the speed of oxen. Speed also increased the range of market transactions among all sectors of urban and rural society. "The rise of horse hauling significantly raised the velocity of goods transportation and hence, in the opposite direction, money circulation, then, this to, would have directly stimulated the economy."[23]

Horses also seem to have been preferred by the peasants of medieval England because of their versatility. They bred them, sold them to wealthier farmers and then bought them back as aging horses for a nominal amount of 2 shillings or less to work for a few more years. "In other words, it was more a part of a survival strategy [to help counter the more stringent human conditions of the so-called 'long thirteenth century' than a calculated plan to improve agricultural production or commercial opportunities."[24]

Nevertheless, survival strategy that it may have been, the adoption of horses as a biological converter of energy for farm work contributed to greater market development for agricultural products and the introduction of vehicular innovations, such as wheeled plows and two-wheeled carts, made of ash. Food for horses also underwent changes with the growing of oats and legumes, increasing the caloric value of their diets that they, in turn, metabolized for work.[25]

Historically, the food they consumed for energy came from pasture grasses and hay, and unlike oxen that required hours for digestion before working, horses' digestive systems allowed them to work soon after eating. Several feedings a day with ample amounts of water made these animals living machines. Although we do not have accurate information about the diets of horses globally, L. H. Bailey's *Cyclopedia of American Agriculture* (1908) reported on detailed experiments with horses pulling omnibuses in Paris and London. There working horses burned 7,902 calories per 500 pounds of weight while horses at rest burned 4,356 calories just to maintain its bodily functions of breathing, heart rate and digestion. For draft horses, "to move one ton 20 miles on a level road at 2.9 mph in one day required 3,421 calories for a 1,100-pound horse."[26]

By the end of the thirteenth century, horses hauled more than 75 percent of the consumable products. Horse-drawn vehicles carried goods to market in growing urban centers. It also increased the speed with which coinage, silver and gold were circulated from cities to towns and vice versa. During winters in which ice-clogged navigable rivers and streams, shod horses delivered the goods. Not until the fifteenth century in medieval England, however, did all-horse plough teams replace teams of oxen and mixed teams of horses and oxen. And as noted earlier, they contributed significantly to the market economy and created closer ties between towns and rural villages. Most important was the contribution of peasants, rather than large landowners, who proved to be the most progressive technologically in the use of horses. Also, they innovated in the growing of legumes for animal and human consumption.[27]

New machinery allowed fledgling mining operations to dig deeper and deeper to extract mineral wealth. An ingenious machine constructed in Hungary and Saxony in the fifteenth century solved the old problem of mine drainage. As miners dug deeper shafts, they eventually encountered water seeping in and often flooding mine cavities. To eliminate water from the deepest pits, it was pumped up in three flights before it reached the surface and was carried away.

> Each pump was set in motion by the rotation of a large horse-driven wheel. To turn these wheels, laborers led horses down inclined shafts, which sloped and twisted like screws. In construction these shafts apparently resembled the ramps that enable modern automobiles owners to park their cars in congested city areas. The work required ninety-six horses. They were employed in relays at each of the three wheels.[28]

The resiliency of horses, a fixture from antiquity until the modern age, maintained its visibility into the twentieth century. Despite the invention of the steam engine and the development of railway networks throughout Europe in the nineteenth century, the ubiquity of the horse remained a significant player in the continent's transportation system. For example, in England, horse-drawn transport filled the gaps left by the railways. In urban areas, it provided short-haul transport for passengers and goods. Accounting for increases in the country's human population, the growth of the horse population not only survived; it also thrived in an increasingly mechanized world. In 1811, there were 251,000 horses; in 1851, 264,000; and in 1901, an astonishing 1,100,000 horses![29] The transition from one energy regime to another remained fluid as older ways of providing power coexisted with newer forms of energy. Human and animal labor continued unabated, and for much of human history, only the ratio changed. In history, minerals (coal, oil and gas) would replace organic sources of energy (humans, animals and wood). It is noteworthy, however, that wood stocks would remain the energy source to power steam engines in factories and steamboats on internal waterways for decades.

In the United States, as late as the early decades of the nineteenth century, farm labor was hand labor with few horses. By the 1830s and 1840s, mechanization on the farm increased the number of horses used. By 1860, rural and urban enterprises owned 7 million horses. By the turn of the century, the number had jumped to 25 million, with 396 horses per square mile in the cities of the Northeast and with 541 per square mile in the Midwest. In 1900, 130,000 horses worked in Manhattan (borough of New York City), the country's largest city and 74,000 in Chicago, then the country's second-largest city.[30]

The nineteenth- and early twentieth-century Western world was one powered by horses. In England, on the European continent and in the United States, the urban horse delivered raw materials to factories and carried away its finished products. Workers loaded them on railcars for shipment to distant places and on ships for travel to other countries. Within cities, horses delivered meat, fish and dairy products, malt and hops to breweries and beer and spirits to saloons. They pulled fire wagons, ambulances and hauled away garbage. Their manure was swept up and loaded onto horse-drawn wagons for delivery to farms as fertilizer. Horses became an integral part of the "organic city." As U.S. cities' populations grew rapidly, with a massive immigration from Europe (1880–1920), this horse-powered world became a mixed blessing. Depending on size, each horse produced between 20 and 50 pounds of manure each day and a gallon of urine.[31] Then, they became known as urban polluters and were soon thereafter replaced by mechanized vehicles, automobiles and trucks. Ironically, gasoline-powered vehicles, a major source of urban pollution, were viewed then as solutions to the feces and urine of the ubiquitous horse.

Its role as a working horse would also diminish by 1945. Yet, as noted, a century before, horses became shackled to a new generation of agricultural inventions. Forty horse-drawn teams became familiar sights pulling harvesters across wheat fields. The number of such horses on the farm dropped perceptibly from about 19 million in 1910 to fewer than 3 million in 1954. They had replaced human workers in the agricultural world and machines with internal combustion engines would replace them as modernization arrived on the farms. The replacement of the urban horse with motorized vehicles completed the transition from one energy regime to another, from human and animal power to gasoline-powered machinery and vehicles. Forgotten was a time when 130,000 working horses in Manhattan pulled carts, wagons and carriages moving food, raw materials for manufacturing and construction and people across space and time.[32]

As mechanized vehicles, tractors, trucks and automobiles replaced horses as alternative sources of power. The continued use of the term *horsepower* in the age of the machine is a constant reminder of the role of the horse in civilization. The term is used to measure the power of an engine:

> It is a measure of the drawing power of a horse. [O]ne horsepower (hp) is equivalent to 746 watts. The term dates from the beginning of the nineteenth century when horses were in widespread use as providers of power for engines and machines, for example for grinding, spinning, and furnace blowing. 1hp could crush 32 bushels of malt per hour in a brewery.[33]

James Watt estimated that a pony was able to lift 220 pounds of coal up a 100-foot shaft in one minute, which equaled 22,000 foot-pounds per minute. A mature horse lifted in the same amount of time 50 percent, or 33,000 foot-pounds a minute. This is the measure used to calculate horsepower for cars, trucks and other mechanized vehicles today.

Once human populations begin to grow, requiring more land for habitation, fertile land, the converter producing food for human and pastures for animals, created an ecological disequilibrium. Either farmers moved to cultivate new agricultural land or they cultivated existing land more intensively. Humans replaced slash and burn agriculture as land became scarcer. Dry farming with rotating crops and irrigation without rotation followed in succeeding centuries.

The wood civilization

Economic historian Paolo Malanima, argues that from an energy use viewpoint, the discovery and use of fire enabled humans to "greatly improve their technical capacities."[34] He refers to the long period before industrialization as the "wood civilization." Wood for fuel in homes for cooking and heating represented about 90 percent of its use in colder regions cut across space and time. For the remainder, humans used trees, an organic substance, to cut into timber at sawmills to build houses (boards, beams, rafters, crossbeams, laths, roofing, window frames), make furniture (chairs, benches, tables, bed frames) and utensils (pots, ladles, and dishes) and tools (ax handles, hoes, scythes), as well as carts and boats.

According to Lewis Mumford,

> wood was the universal material of the geotechnic economy of the pre-industrial world. As for the common tools and utensils of the time, they were more often of wood than of any other material. The carpenter's tools were of wood, the rake, the ox yoke, the cart, the wagon, were of wood: so was the washtub in the bathroom: so was the bucket and the broom: so in certain parts of Europe the poor man's shoe. Wood served the farmer and the textile worker: the loom and the spinning wheel, the oil presses and the wine presses were of wood, and even a hundred years after the printing press was invented, it was still made of wood. The very pipes that carried water into the cities were often tree trunks: so were the cylinders and the pumps. One rocked a wooden cradle; one slept in a wooden bed; and when one dined one "boarded." One brewed beer in a wooden vat and put liquor in a wooden barrel. Stoppers of cork, introduced after the invention of the glass bottle, began to be mentioned after the fifteenth century.
>
> The ships of course were made of wood and pegged together with wood; but to say that is only to say that the principal machines of industry were likewise made of wood: the lathe, the most important machine tool of the period, was made entirely of wood – not merely the base but the movable parts. Every part of the windmill and the water-mill except for the grinding and cutting elements was made of wood, even the gearing: the pumps were chiefly of wood, and even the steam engine, down to the nineteenth century, had a large number of wooden parts: the boiler itself might be of barrel construction, the metal being confined to the part exposed to the fire.[35]

Despite the differences in climate across Eurasia, dense forests were punctuated with small settlements of humans. Wood's availability for heating and cooking within close proximity to early cities supported their growth. As an energy source, the human energy and horsepower required to hew and haul timber meant that it had to be located on the periphery of cities. Much of its weight in water limited its efficiency as a fuel. Its high carbon content, when compared to fossil fuels, meant that burning wood added carbon dioxide to the atmosphere's chemical composition. Transforming wood to charcoal intensified the air's carbon content and led to the exploitation and overharvesting of a region's forests.[36]

Rome's use of wood as a fuel

In antiquity, Romans used wood for fuel (heating and cooking) and for construction material, consuming much of the surrounding forests with little regard for conservation. The heating of the popular Roman public baths by 100 CE to temperatures uncommon by modern standards required so much wood that forests were designated for their use only. In experiments conducted at a surviving bath site many centuries later, it was estimated that a site smaller than a public one required 114 tons of firewood each year to keep bath temperatures at 130 degrees Fahrenheit and a sweating room at 160 degrees Fahrenheit.[37]

Some historians have argued that the depletion of woodlands represented one of the causes for the Roman Empire's collapse from assaulting Germanic tribes. As the only fuel source mostly, depletion caused the prices for wood and wood products to escalate, precipitating ruinous inflation. Inflation caused conflicts, some of them bloody, among Rome's military legions competing for wood for chariots and weapons. With deforestation, large areas of the empire experienced changes in their microclimates. Woodlands that once protected the ground from intense heat in the summer disappeared, and the ravages of winter winds sweeping across a denuded landscape blew away its precious soil. Some fertile agricultural lands became deserts. Escalating food prices hit nonelites hardest, further weakening Rome's political and economic stability.

Wood fuel in medieval and early modern Europe

With the fall of Rome in 476 CE, the Western Empire experienced a long economic and societal depression. Declining populations saw some cities reverting to small towns. Rome itself experienced a dwindling population. An overall depression relived the pressure on forests for fuel and timber, resulting in a long period of natural reforestation. In Europe, with a population of fewer than 30 million in 900 CE, fewer than three people inhabited an area of 1 square kilometer or 0.387 miles. Much of which was a wooded landscape. Within two centuries, according to the Domesday Book (1086 CE), 15 percent of England's landmass was covered in forests. When growth and development returned two centuries later, the ax had cleared between one-third and one-half of the country's woodlands. In the centuries after 900 CE, land was needed to grow crops for humans and fodder for draft animals. The attack on the forests began across the continent with urban population growth. Given Europe's seasonal temperatures, winter months forced farmers into the woods to cut trees for firewood to heat their homes and cook their food. The human use of the ax cut back the forest's canopy, exposing it to solar radiation, agriculture's traditional energy source.

From 1300 CE onward, shortages of firewood caused inflated prices along with increased transportation costs. Cheaper fuel sources, such as sea coal in England, mined near Newcastle and brought to London by ship, and peat from the Netherlands offered another energy source. The use of coal, however, limited as it was in the fourteenth century marked the transition from an overall dependence on organic energy sources to a progressive dependence on fossil fuels. Between 1450 and 1700 CE, Londoners experienced the gradual depletion of woodlands on its periphery. In the 1600s, few wood lots in England were greater than 20 acres, and in southeastern England, there were few woodlots at all. The price of firewood was double the rising rate of commodity prices. During this same period, the city's population surged from 45,000 to 530,000. At 45,000, the city required about 1,350 cords of firewood each year. (A stacked cord of wood is 8 feet long, 4 feet wide and 4 feet deep).

By comparison, in Paris, France in 1800, residents who could afford it paid as much as 10 percent of their annual income for firewood cut as much as 100 miles away. By this time, European households competed with the growing iron industry that needed charcoal for its smelters. Iron making required 14 tons of charcoal to smelt one ton of iron in the early decades of the nineteenth century. In the decade between 1830 and 1840, European and U.S. ironworks companies owned more than 8,000 acres of forests. Such a large area was necessary because iron making of this magnitude required 6,000 cords of wood (160,000 bushels of charcoal) to produce 1,000 tons of iron.[38]

Compared with European households that burned approximately 15 cords of wood each year, residents in the United States, where large regions of virgin forests existed, burned forty cords per family annually. In the first century of European settlement of North America in the seventeenth century until the 1930s in the United States, settlers and then citizens burned 12.5 billion cords of wood for industrial and domestic uses. By 1850, wood provided more than 90 percent of the energy consumed in the United States. Not until 1960 did it drop to its lowest level of 5 percent, where it remains today.[39]

Until the invention and use of cast-iron stoves, most of the heat generated by the open fireplace disappeared up the chimney, with only 5 to 10 percent heating a small area. Stoves led to energy efficiencies of up to 15 percent, but in warmer Southern Europe, open fireplaces remained commonplace until later in 1800.

> Since calories per kilogram (kg) of dry firewood are more or less 3000–3500, and since people in Europe need an average of at least 2kg of firewood per day per person, in a per capita budget of calories, the ones from wood easily exceed the ones from food or from fodder.[40]

Food, fodder and firewood became essential ingredients for a sustainable living standard in the world before industrialization.

The use of dry firewood was used primarily for heating, yet cooking and illumination were essential for the growth of small manufacturing. Baking wheat bread, a staple in the diets of rural and urban households, required ample supplies of firewood. Pottery makers converted clay into useful pots; brewers and distillers needed firewood to make their beer and spirits. Charcoal, a by-product of burning wood, provided the

> [f]uel for the kilns represented 60 percent of the total cost of the production of bricks, tiles and lime. . . . The ratio between charcoal used and iron obtained was 16 to 1. [And] it was necessary to process 200kgs of mineral and burn 25 square metres of wood to obtain 50 kgs of iron.[41]

Despite the applications of firewood and charcoal to these manufacturing processes, combustion for heating rather than productive purposes remained the norm until the late eighteenth and the entire nineteenth century in England and throughout much of Europe. Before the invention and utilization of the "iron horse" (locomotives) for transportation, craftspeople used wood to build chariots, carts and wagons. For the weapons of war, wood for lances, spikes, ax handles, war wagons, wheels and catapults consumed thousands of acres of woodlands.

Was there a timber crisis in premodern Europe?

Economic historian Paul Warde has provided a detailed accounting of the "wooden world," using data from medieval Germany but generalizing to the entire Holy Roman Empire. There, firewood used for domestic fuel accounted for as much as 90 percent of the total and remained at the level of 70 percent until the 1850s. However, we know little about its use for either heating or cooking, even though these are not mutually exclusive activities. For example, in the sixteenth century, German homes were designed with many rooms; a kitchen range had a tile stove attached to it that distributed heat to one or more adjoining rooms. How often heat was needed to warm these homes depended on the weather and the length of the winter season, which varied each year. In many of these homes, smoke was sent by flues into the attic to kill off vermin and prevent the accumulation of mold.

Wood for fencing protected large tracts of private land for cultivation and prevented intrusions from grazing animals. This barrier distinguished private holdings from the commons that set aside plots for gardens and orchards used by townsfolk. In 1673, six healthy oak trees provided 900 posts for 7- to 8-foot-high fencing that would have supposedly required hundreds of trees to complete. Given the lengthy life of such fences, their toll on the woodlands would be relatively small.

According to Professor Warde, the demand for vineyard poles, measuring 7 feet long but an inch square, to stake the vines required much more wood stock. For example, "Bietigheim bought 200,000 poles *per annum* around 1570. The two Ingersheims received some 86 cartloads of wood for the 36 hectares of vineland leased by the Duke of Wurttemberg there."[42] According to his calculations, for the woodlands of *Forstamt* Leonberg, 45,000 cubic meters was used for domestic firewood, 15,000 cubic meters for vineland stakes and 10,000 cubic meters for bakers, tile makers, barber-surgeons and blacksmiths.

Wood yields began to decline because of the practices described earlier and by 1760 CE "reached extraordinarily low levels and timber for importing zones along the Rhine, stripped bare large parts of the uplands. Transportation costs remained expensive because of the distances traveled over very poor roads and the enormous stress placed on the horses and oxen as well as on the wagons and carts that they pulled. These stresses forced owners to incur the additional costs of repair and replacement. However, wood would remain the major source of thermal energy for most of Europe. Only England began to shift to coal while the Dutch continued its long-standing use of peat."[43]

The decline of timber in England (specifically London and its periphery) noted earlier occurred centuries before the formal recognition of industrialization. Coal from Newcastle and elsewhere provided the country's growing economy with an energy boost. Also, it served as the transition from the energy constraints of the medieval world to the fossil fuel wealth of the modern world. By 1700 CE, English mining produced 3 million tons of coal per year, with nearly 50 percent used for the domestic market of heating and cooking. Manufacturing use accounted for almost 40 percent, with the smelting and refining of metals (except iron, which would come later), brickmaking, salt processing, dyeing, blacksmith work, malting, brewing, distilling, glass blowing and ceramics making up the majority of uses.[44]

Regional deforestation led some historians to conclude that its environmental impact created a "timber crisis" in England specifically and Europe more generally. In their view, the early modern economy faced growth limitations because of its energy dependence on agriculture, animals, forests and the limited energy provided by water and wind. Population growth and economic expansion in England caused wood scarcity. Its growing iron industry, a major consumer of wood, reached an economic bottleneck. The transition to large quantities of fossil coal broke through this "ecological crisis" and made it possible to restructure England and Europe's economy. Mining coal and inventing the steam engine were causes for the Industrial Revolution of the eighteenth century. The validity of their conclusion that a timber shortage caused the transition to coal follows.

To test this conclusion that a "timber crisis" was a European-wide phenomenon, causing England to begin using coal two centuries before industrialization, economic historian Robert C. Allen studied fuel prices for the medieval and early modern periods. They included prices for coal, peat, charcoal, cords of hardwood and softwood in London and its environs, Newcastle, Amsterdam and other Dutch cities, Antwerp, Strasbourg, Paris, Milan, Florence, Naples, Valencia, Madrid, Leipzig, Vienna, Gdansk and Warsaw.[45] He concluded that despite wood scarcity in greater London and some cities in Belgium and the Netherlands, one could not generalize for the countries in question or to the entire continent.

In London, firewood prices rose rapidly in the late sixteenth century exceeding the price of coal. As Professor Warde noted for Germany, transportation costs (i.e., poor roads, wagon and cart repairs, and horses) inflated wood prices. "Wood doubled its price in three miles,"[46] creating regional markets for wood based on the proximity of the source to its end users. To exacerbate wood's rising price, coal sold at a 50 percent discount of wood, even though consumers viewed it as an inferior fuel source because of its smell and the impurities it left in finished products.

Did peat fuel the Dutch economic miracle?

Regarding energy consumption in the Netherlands, particularly the province of Holland, the Dutch Golden Age (1500–1670 CE), the traditional account by the Dutch historian J. W. de Zeeuw argues that without the cheap energy fuel, peat, the economic miracle of the seventeenth century would not have been possible.[47] A counterargument argues for a more limited role for peat in explaining the Dutch economic miracle during the late medieval period. Beginning in 1350

CE, Holland faced an ecological crisis with declining agricultural productivity caused by the erosion of peat soils and the rise of internal water levels. Reclaiming wetlands for agriculture from 900 to 1300 CE contributed to the crisis by causing gradual land subsidence and saltwater incursions, especially during the winter months.

In response, peasants shifted planting from winter grains of rye and wheat, both bread grains, to summer grains, barley and oats, both important for the growing brewery industry. In addition, they switched to cattle farming, peat digging and fishing. Begun on the inland lakes but spreading to the Zuiderzee and the North Sea, this change diversified the employment of Dutch peasants and linked their products to the growing urban market at home and abroad. By 1500 CE, only 20 percent of the Dutch labor force worked in agriculture. However, large economic outputs in the brewing of beer, textile manufacturing, the herring catch, and services made Holland's industry, fisheries and shipping competitive on the international market. Despite rising grain prices from Northern France and the Baltic states, increased wages for Dutch laborers gave them the means to pay for their food, a form of energy. The gains in brewing, clothing, digging, cattle farming and the shipping industry more than compensated for the losses in agriculture. The Dutch solved their chronic food shortages by concentrating on their exports in products and services.

The digging of peat, a cheap, plentiful and nonrenewable energy source, contributed to Holland's economic expansion. Some export industries depended on it. Brewing beer, brickmaking and salt production were energy-intensive activities. Did this relatively cheap source of energy give the Dutch a comparative advantage over other nations, as much of the literature suggests. According to historian Jan Luiten van Zanden,

> [p]er unit of energy peat in Leiden was probably more expensive than coal in southern England, the difference amounted to about 25% in 1620–1639. During the early modern period per capita energy consumption in Holland was also less than in England, which also points to the fact the possession of peat did not give rise to significant comparative advantages.[48]

Digging peat posed environmental costs. As a nonrenewable energy source, it scared Holland, a province whose elevation at and below sea level was made more vulnerable by creating more wetlands and inland lakes. Digging peat in proximity to its distilleries, breweries, soap-boiling, sugar-refining and brick- and dye-making facilities limited production and transportation costs. Once depleted, digging commenced in more distant areas, raising costs. As a result, imported firewood and coal became competitive sources of energy. The explosive growth of firewood consumption in Amsterdam is a case in point. From 1730 to 1760, the city's residents and manufactories burned 72,000 feet of firewood initially and later between 150,000 and 180,000 feet as fuel. In the industries listed earlier, imported coal dominated the energy market.[49] Claims that plentiful and cheap peat explain the Dutch Golden Age without acknowledging the availability of imported food, as a source of calories and imported wood and coal as sources of mechanical energy are mistaken.

Conclusion

Food for humans and fodder for animals remained primary fuel sources for most of human history. Fire, burning biomass, animal dung, straw and wood have been, and for much of the undeveloped world remain, primary sources for heating and cooking into the twenty-first century. In developing countries, dependency on wood for fuel is commonplace. In Africa, this represents 90 percent of the total amount harvested. In the millennia before industrialization, wood held a commanding lead over other fuels. With coal used marginally from the 1100s CE, its growth coincided with the rising costs of wood for heating, cooking and charcoal making used for the smelting of metals such as iron, copper and lead. When the English entrepreneur Abraham Darby produced coking coal as a substitute for charcoal in making iron in 1709 CE, it forecast the ultimate demise of charcoal for manufacturing purposes. At the same time, however, wood remained the fuel of choice for the network of steam-powered locomotives and the construction of the wood railways and bridges that linked urban and rural communities until coal prices became more competitive. By 1860, the railroads in the United States burned 6 million cords of wood, while the steamboats consumed 3 million cords.[50]

Rising wood-fuel prices in Europe occurred at a time when early settlers to North America encountered virgin forests that spread across the continent. As the new nation of the late eighteenth and nineteenth centuries began to grow, the United States became a leader in the application of advanced timber cutting and removable technology. Human energy remained the primary engine of lumberjacks using the ax and crosscut saw. Humans and their horses pulled cut trees to rivers to flow them downstream to water-powered sawmills. Within a few decades of the 1900s CE, steam-powered winches and pulleys did the work of men and horses. Gasoline-powered chain saws and motorized vehicles followed in the twentieth century. The timber became lumber for the growing construction industry, packaging in the form of boxes, paper bags and cardboard for multiple purposes.

The Germans invented the mechanical process pulping of wood in 1844. Americans in Pennsylvania commercialized the process in 1867. Along with soda pulping and sulfite processing, wood pulp replaced rags in papermaking with another German invention, Kraft process for pulping wood chips and wood residue into cellulose used for making paper. Literacy, previously limited to elites who could afford to buy books, now became a prerequisite for upward mobility for school-attending workers, immigrants and their kin. Between 1870 and 1920, paperback books and newsprint, as well as packaging materials, used 7.2 million tons of pulp, and by 1960, paper production using the Kraft Process reached 40 million tons annually.[51]

Steam-powered machinery and gasoline-powered vehicles replaced horses on the farm and in the city. Most disappeared from the rural and urban landscape by the early decades of the twentieth century. Their presence as warhorses continued until mid-century. They remained fixtures on parade routes, at thoroughbred racetracks and steeplechases and as mounts for the policing of crowds. Their presence today recalls a time when horses became substitutes for human laborers.

Notes

1 Paolo Malanima, "Energy Consumption in the Roman World," in *The Ancient Mediterranean Environment between Science and History*, ed. W.V. Harris (Boston: Brill, 2013), 15.
2 Paolo Malanima, *Pre-Modern European Economy: One Thousand Years (10th-19th Centuries* (Leiden: Brill, 2009), 51.
3 Juli G. Pausas and Jon E. Keeley, "A Burning Story: The Role of Fire in the History of Life," *BioScience* 59, no. 7 (July/August 2009): 597.
4 Kristen Hawkes, "The Grandmother Effect," *Nature* 428 (2004): 128–129.
5 E.A. Wrigley, "Energy Constraints and Pre-Industrial Economies," in *Economia e Energia, SECC.XIII-XVIII, Instituto Internazionale Di Storia Economica*, ed. Simonetta Cavaciocchi (Le Monnier, 2002): 156–157.
6 Jared Diamond, *Guns, Germs and Steel: The Fates of Human Societies* (New York: W.W. Norton and Co., 1997), 42.
7 Ibid.,75–76, 80.
8 E.A. Wrigley, "Energy Constraints and Pre-Industrial Economies," in *Economia e Energia*, 158–159.
9 Orjan Wikander, "Sources of Energy and Exploitation of Power," in *The Oxford Handbook of Engineering and Technology In The Classical World*, ed. John Peter Oleson (New York: Oxford University Press, 2008), 136.
10 N. Koepke and J. Baten, "Agricultural Specialization and Height in Ancient and Medieval Europe," *Explorations in Economic History* 45, no. 2 (April 2008): 142.
11 David W. Anthony, "Bridling Horse Power: The Domestication of the Horse," in *Horses Through Time*, ed. Sandra L. Olsen (Boulder, CO: Roberts Rinehart Publishers, 1995), 63.
12 Clay McShane and Joel A. Tarr, *The Horse in the City: Living Machines in the Nineteenth City* (Baltimore, MD: Johns Hopkins University Press, 2007), 7.
13 Anthony, "Bridling Horse Power," 59.
14 Juliet Clutton-Brock, *Horse Power: A History of the Horse and the Donkey in Human Societies* (Cambridge: Harvard University Press, 1992), 168–169.
15 Herodotus, The Histories IV, 46–47 as quoted in Anthony, "Bridling Horse Power," 60.
16 As quoted in, H.G. Creel, "The Role of the Horse in Chinese History," *The American Historical Review* LXX, no. 3 (April 1965): 667.
17 Verlyn Klinkenborg, "A Horse is a Horse, of Course," A Review of Ulrich Raulff, "Farewell to the Horse: A Cultural History," *The New York Review of Books*, LXV, no. 3 (February 22, 2018): 46.
18 Paolo Malanima, "Energy Systems in Agrarian Societies: The European Deviation," in *Economia e Energia*, 74.
19 Ibid.
20 Ibid.
21 John Langdon, "The Uses of Animal Power from 1200 to 1800," in *Economia e Energia*, 214.
22 Ibid., 215.
23 John Langdon, *Horses, Oxen and Technological Innovation: The Use of Draught Animals in English Farming from 1066 to 1500* (Cambridge, UK: Cambridge University Press, 1986), 272.
24 Langdon, "The Uses of Animal Power," 216.

25 Ibid., 219.
26 W.H. Jordan, "The Feeding of Animals," in *Cyclopedia of American Agriculture: A Popular Survey of Agricultural Conditions, Practices, and Ideals in the United States and Canada*, 4 vols., ed. L.H. Bailey (New York: Macmillan, 1908), 3, 78–87. As reported in McShane and Tarr, *The Horse in the City*, 127.
27 John Langdon, "Horses, Oxen and Technological Innovation," 291–292.
28 Georgius Agricola. *De re metallica* (Hoover ed.; London, 1912), 1994–195. As cited in John Nef, *The Conquest of the Material World* (Chicago: Chicago University Press, 1964).
29 T.C. Barker, "Transport: The Survival of the Old beside the New," in *The First Industrial Revolutions*, eds. Peter Mathias and John A. Davis (Oxford: Basil Blackwell, 1989), 100.
30 Ann Norton Greene, *Horses at Work: Harnessing Power in Industrial America* (Cambridge, MA: Harvard University Press, 2008), 171.
31 McShane and Tarr, *The Horse in the City*, 16.
32 Klinkenborg, "A Horse is a Horse, of Course," 47.
33 Clutton-Brock, *Horse Power.*
34 Malanima, *Pre-Modern European Economy*, 50.
35 Lewis Mumford, *Technics and Civilization* (New York: Harcourt, Brace & Co., 1934), 119–120.
36 Luc Gagnon, "Civilisation and Energy Payback," *Energy Policy* (Elsevier, 2008): 3318.
37 John Perrin, *A Forest Journey: The Role of Wood in the Development of Civilization* (Cambridge, MA: Harvard University Press, 1991), 112.
38 David A. Tillman, *Wood as an Energy Resource* (New York: Academic Press, 1978), 8.
39 Harvey Green, *Wood: Craft, Culture, History* (New York: Penguin Books, 2006), 362–364.
40 Ibid., 74–75.
41 Ibid., 58–59.
42 Paul Warde, *Ecology, Economy and State Formation in Early Modern Germany* (Cambridge, UK: Cambridge University Press, 2006), 269–270.
43 Ibid., 357.
44 Peter Mathias, "Economic Expansion, Energy Resources and Technical Change in the Eighteenth Century: A New Dynamic in Britain," in *Economia e Energia*, 28.
45 Robert C. Allen, "Was There a Timber Crisis in Early Modern Europe?," in *Economia e Energia*, 469–481.
46 G. Hammersley, "Crown Woods and their Exploitation in the Sixteenth and Seventeenth Centuries," *Bulletin of the Institute of Historical Research* 30 (1957): 157.
47 J.W. de Zeeuw, "Peat and the Dutch Golden Age: The Historical Meaning of Energy-Attainability," A.A.G. *Bijdragen* 21 (Wageningen, 1978), 3–31.
48 Jan Luiten van Zanden, "The Ecological Constraints of the Early Modern Economy: The Case of the Holland 1350–1800," in *Economia e Energia*, 1027.
49 Ibid., 454–455.
50 Tillman, *Wood as an Energy Resource*, 11.
51 Ibid., 12, 16.

References

Agricola, Georgius. *De Re Metallica*. London: Mining Magazine, 1912.

Allen, Robert C. "Was There a Timber Crisis in Early Modern Europe?" In *Economia e Energia, SECC.XIII-XVIII, Instituto Internazionale Di Storia Economica*, edited by Simonetta Cavaciocchi, 469–482. Firenze, 2002.

Anthony, David W. "Bridling Horse Power: The Domestication of the Horse." In *Horses Through Time*, edited by Sandra L. Olsen, 57–82. Boulder, CO: Roberts Rinehart Publishers, 1995.

Barker, T.C. "Transport: The Survival of the Old Beside the New." In *The First Industrial Revolutions*, edited by Peter Mathias and John A. Davis. Oxford: Basil Blackwell, 1989.

Clutton-Brock, Juliet. *Horse Power: A History of the Horse and the Donkey in Human Societies*. Cambridge, MA: Harvard University Press, 1992.

Creel, H.G. "The Role of the Horse in Chinese History." *The American Historical Review* LXX, no. 3 (April 1965): 647–672. doi: 10.2307/1845936.

de Zeeuw, J.W. "Peat and the Dutch Golden Age: The Historical Meaning of Energy-Attainability." A.A.G. *Bijdragen* 21 (Wageningen, 1978): 3–31.

Diamond, Jared. *Guns, Germs and Steel: The Fates of Human Societies*. New York: W.W. Norton and Co., 1997.

Gagnon, Luc. "Civilisation and Energy Payback." *Energy Policy* 36, no. 9 (September 2008): 3317–3322. doi: 10.1016/j.enpol.2008.05.012.

Green, Harvey. *Wood: Craft, Culture, History*. New York: Penguin Books, 2006.

Greene, Ann Norton. *Horses at Work: Harnessing Power in Industrial America*. Cambridge, MA: Harvard University Press, 2008.

Hammersley, G. "Crown Woods and their Exploitation in the Sixteenth and Seventeenth Centuries." *Bulletin of the Institute of Historical Research* 30 (1957): 138–157.

Hawkes, K. "The Grandmother Effect." *Nature* 428 (March 2004): 128–129. doi: 10.1038/428128a.

Jordan, W.H. "The Feeding of Animals." In *Cyclopedia of American Agriculture: A Popular Survey of Agricultural Conditions, Practices, and Ideals in the United States and Canada*, edited by L.H. Bailey. New York: Macmillan, 1908.

Klinkenborg, Verlyn. "A Horse Is a Horse, of Course," A Review of Ulrich Raulff, "Farewell to the Horse: A Cultural History." *The New York Review of Books* LXV, no. 3 (February 22, 2018): 46.

Koepke, N., and J. Baten. "Agricultural Specialization and Height in Ancient and Medieval Europe." *Explorations in Economic History* 45, no. 2 (April 2008): 127–146. doi: 10.1016/j.eeh.2007.09.003.

Langdon, John. *Horses, Oxen and Technological Innovation: The Use of Draught Animals in English Farming from 1066 to 1500*. Cambridge, UK: Cambridge University Press, 1986.

Langdon, John. "The Uses of Animal Power from 1200 to 1800." In *Economia e Energia, SECC.XIII-XVIII, Instituto Internazionale Di Storia Economica*, edited by Simonetta Cavaciocchi, 213–222. Firenze, 2002.

Malanima, Paolo. "Energy Systems in Agrarian Societies: The European Deviation." In *Economia e Energia, SECC.XIII-XVIII, Instituto Internazionale Di Storia Economica*, edited by Simonetta Cavaciocchi, 61–99. Firenze, 2002.

Malanima, Paolo. *Pre-Modern European Economy: One Thousand Years (10th-19th Centuries)*. Leiden: Brill, 2009.

Malanima, Paolo. "Energy Consumption in the Roman World." In *The Ancient Mediterranean Environment between Science and History*, edited by W.V. Harris, 13–37. Boston: Brill, 2013.

Mathias, Peter. "Economic Expansion, Energy Resources and Technical Change in the Eighteenth Century: A New Dynamic in Britain." In *Economia e Energia, SECC.XIII-XVIII, Instituto Internazionale Di Storia Economica*, edited by Simonetta Cavaciocchi. Firenze, 2002.

McShane, Clay, and Joel A. Tarr. *The Horse in the City: Living Machines in the Nineteenth City*. Baltimore, MD: Johns Hopkins University Press, 2007.

Mumford, Lewis. *Technics and Civilization*. New York: Harcourt, Brace & Co., 1934.

Nef, John. *The Conquest of the Material World*. Chicago: Chicago University Press, 1964.

Pausas, Juli G., and Jon E. Keeley. "A Burning Story: The Role of Fire in the History of Life." *BioScience* 59, no. 7 (July/August 2009): 593–610. doi: 10.1525/bio.2009.59.7.10.

Perrin, John. *A Forest Journey: The Role of Wood in the Development of Civilization*. Cambridge, MA: Harvard University Press, 1991.

Tillman, David A. *Wood as an Energy Source*. New York: Academic Press, 1978.

van Zanden, Jan Luiten. "The Ecological Constraints of the Early Modern Economy: The Case of the Holland 1350–1800." In *Economia e Energia, SECC.XIII-XVIII, Instituto Internazionale Di Storia Economica*, edited by Simonetta Cavaciocchi, 1011–1030. Firenze, 2002.

Warde, Paul. *Ecology, Economy and State Formation in Early Modern Germany*. Cambridge, UK: Cambridge University Press, 2006.

Wikander, Orjan. "Sources of Energy and Exploitation of Power." In *The Oxford Handbook of Engineering and Technology In The Classical World*, edited by John Peter Oleson, 136–157. New York: Oxford University Press, 2008.

Wrigley, E.A. "Energy Constraints and Pre-Industrial Economies." In *Economia e Energia, SECC.XIII-XVIII, Instituto Internazionale Di Storia Economica*, edited by Simonetta Cavaciocchi, 155–172. Firenze, 2002.

2 Early uses of wind and waterpower

Introduction

All energy is either the direct or indirect capture of solar energy. Chapter 1 focused on the organic economy of human and animal muscle and wood. In the case of humans and animal muscle, some movement is possible but at great expense and with diminishing returns. The energy derived from waterwheels is restricted to riverbanks while the power of wind when applied to sails becomes highly movable but when applied to mills remains stationary.

Windmills and watermills operate on the same principle. Motion means kinetic energy, whether it's air or water in motion. The particles of air are gaseous while those in water are liquid. Capturing the movement of one in a windmill or the other in a watermill means capturing their energy and harnessing it to do work. Wind and water are forms of solar energy. For wind, it is the heating of the atmosphere by the sun. For water, it is the planet's rotation and the irregularities of Earth's surface. Energy from wind and water replaced humans and animals from the drudgery and monotony of repetitive activity. Never entirely, but their use declined rapidly during the modern era.

In terms of capital investment, harnessing the power of water required the construction of dams to contain flowing water and create holding ponds and reservoirs. The dredging of shallow impediments, the digging and maintaining of canals and sluices, the construction of floodgates and locks added to the costs. Mill construction consumed large quantities of lumber for framing, flooring, siding as well as a mill's basic machines of spindles and looms. Lumber with some iron reinforcement represented the primary material for the wheelhouse, the wheel and buckets or paddles and gears. Large quantities of cut timber held riverbanks and canal sidings in place.

Yet, despite the size of waterwheels and the more recent efficient water turbines, the river, the source of power, was neither transportable nor dependable. Despite dams built to create holding ponds, streamflow ultimately depended on the size of upstream snowpacks and/or rainfall. Unpredictable mild winters and droughts interfered with a mill's productivity. Higher-capacity watermills combined with new windmills resulted in an expansion of grain grinding capability. The increased milling potential of grain into flour became noteworthy. In terms

Figure 2.1 Constructed 400 years BCE in the city of Hama, Syria, ancient norias (wheels of pots) lifted water from the moving Orontes River and delivered it to aqueducts to irrigate gardens and fields and for household use.

of the total mills available in 1300 CE, the figure of 10,000 seems like a reasonable estimate. The figure of 15,000 is speculative, although historian Richard Holt in *The Mills of Medieval England* regards 12,000 as an acceptable figure for the country in 1300 CE.

During the bubonic plague years (1349–1351 CE), in which hundreds of thousands died in England, windmill use declined precipitously. Before the plague, such figures suggest that a substantial number of English men, women and children worked in the milling industry and gained their wages by operating and maintaining the mills. As many as 50,000 to 100,000 persons got their flour from the approximately 15,000 employed in water- and wind-milling activities. For perspective, hand mills operated by peasant labor did 20 percent of milling activity. By 1540 CE, however, only about 45 percent of those mills survived as peasant populations declined and watermills became more productive. A population rebound in the later sixteenth and early seventeenth centuries led to the construction of more windmills to relieve the pressure on watermills. Owners of watermills contributed to the realignment by pushing for a diversified use of water and wind power.[1]

Waterpower in ancient Rome

In addition to the many uses of wind power, watermills achieved prominence in the ancient world as "one of the first kinds of machine to employ non-human and renewable forms of energy as their motive power [and] amongst the first machines to be automated and some of the first to be used in factory-scale production."[2] This conclusion runs counter to a position held previously by historians of technology, who, from approximately 1935 to 1975, denied the importance of waterpower in the ancient world. They promoted the "technological stagnation thesis," arguing that the watermill was known in the ancient world but seldom used for productive purposes. Its general use, they argued, would await the technological revolution of Europe's medieval period.[3] The historian Marc Bloch, the initial proponent of this thesis, argued that

> [t]he lords of the great *latifundia* who were much less sympathetic to the burdens of humble people, had no reason to install expensive machinery when their markets and their very houses were overflowing with human cattle. As for the more modest households and for bakers, who would in any case have been unable to afford such heavy expenditure, many of them were quite well enough off to have their own domestic slaves; or else they did their own work themselves.[4]

In recent years, historians have reexamined this conclusion and rejected it.[5] In the last 20 years, both archeologists and historians of the Roman Empire have identified more than 56 vertical waterwheel sites from 150 to 500 CE. For the owners and supervisors of some industries, waterwheels, rather than human or animal power, were the preferred technology. However, with as much as one-third of the population held in slavery and many more as poor peasants, work, the expenditure of energy, resulted from human labor, in many cases, using hand mills to grind grain. The widespread use of donkey-powered mills supplemented human labor. Regarding the earliest dates for the use of waterwheels, mill race timbers dated by dendrochronology to 57/58 CE. Pottery remains and coins dated its abandonment about 80 CE.[6]

The existence of water-powered devices for lifting water out of irrigation channels and spread across farms predated the Roman era. Various water-lifting wheels appeared each functioning differently 200 to 300 years BCE in Egypt during the Hellenistic (Greek) expansion. Later, they appeared as the Roman Empire spread into the Middle East. One wheel contained a rim, making it a treadmill powered by the force of moving men.

Another used the power of donkeys and/or horses turning a capstan connected to the wheel by right-angled gearing. A chain of pots for lifting water would also be used and powered by men on a treadmill or by animals using the familiar right-angled gearing. Romans disseminated all these methods for irrigation and for drainage by conquest. These earlier examples point to the continuing use of human and animal power to activate water-lifting wheels, as well as the use of renewable forms of mechanical energy generated by waterwheels.

The massive waterwheel complex at Barbegal, near Arles in southern Gaul (modern France), originally dated from 500 CE is dated now from 300 CE. Excavated in 1937–1938 and fed water from Roman aqueducts, it contained sixteen overshot vertical wheels arranged in pairs on eight terraces. Each wheel measured about 7.3 feet across (2.2 m) and contained two millraces. Its size and complexity represented a breakthrough in watermill technology. It

> may have been a technology of power generation only a century or so old: the overshot vertical mill, where water is channeled into bucket-like containers built into the wheels['] circumference, so that the weight of the captive water (rather than its impact) forces the wheel to turn. Then, via appropriate gearing, the wheel's rotation drove the upper millstone of a rotary quern.[7]

Two stone wheels were used, one resting on the other for grinding. It could deliver the equivalent of about 8 horsepower of mechanical energy to the machinery for grinding grain into flour.

Recent estimates suggest that the Barbegal mills operated at 50 percent capacity. Declines in water flow caused much of the downtime. At 50 percent capacity, they milled 4.5 tons of flour each day, enough to provide 350 grams (12.3 oz) of bread to a population of 12,500 in nearby Arles. It, a city on the rise as an imperial capital in Gaul (modern France) fed a phalanx of bureaucrats, soldiers and plebeians at government expense. Since mostly grains provided carbohydrates, some scholars suggest that 350 grams daily was probably too low for active people:

> Between 200 BCE and 50 CE, the daily per capita rations of grain in Rome varied from about 600 to 1,090 grams, with an average of 600 grams being the absolute minimum for the entire population of Rome, including children and old people.[8]

Figure 2.2 The immense Roman flour mill at Barbegal powered by 16 overshot waterwheels arranged in two parallel rows of 8.

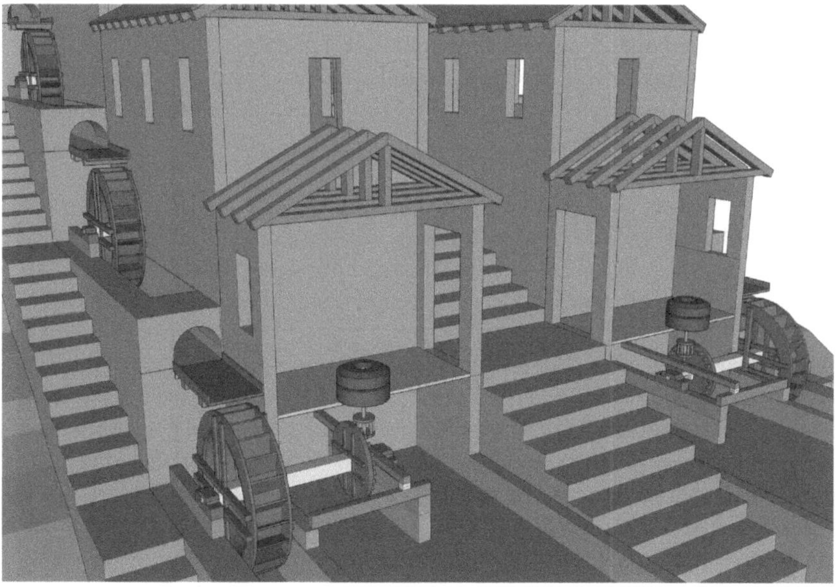

Figure 2.3 One of 16 turbines that transmits power to grindstones (not shown) for milling flour.

The commercialization of bakeries throughout the Roman Empire coincided with a growing urban population. Rome's population reached about 1 million in 1 CE. These figures suggest an output of about 24 kilograms of flour each hour, meaning that the millstones rotated at about 30 revolutions per minute using the power of running water. An oxen, horse, or donkey, would be hard-pressed to sustain revolutions per minute greater than six.

Mill maintenance and food for operators diminished profits. The initial capital costs for Roman mills would be expensive given their wooden construction. The cost of maintenance and repairs of wooden mills impacted their cost/benefit ratio:

> The approximate correspondence between the output estimates for Roman mills and known figures for twentieth century water-mills suggests that the major advances in water-powered milling technology had already been made by the Roman period, and that developments in the last 1,800 years of water-milling largely constituted refinements of an already mature technology.[9]

Large-scale Roman watermills represented a large capital investment to produce a high output of flour. An equally large investment in baking ovens may have been related to the expansion of *annonae*. By government edict, it provided handouts of bread loaves rather than grain to Roman soldiers and plebeians as a way of maintaining political and social stability. It is possible that the army served as an agent for spreading this technology across the empire. Two and possibly three watermill sites have been located along Emperor Hadrian's wall that served as a barrier separating Roman Britain from independent Scotland.

Roman watermills also powered sawmills. Ausonius (ca. 310–ca. 395), the Gallic teacher, described his travels along the Moselle River and the work of a sawmill in his poem *Mosella:*

> Renowned is Celbis [Kyll] for glorious fish,
> And that other [Ruwar] as he turns his millstones
> In furious revolutions and drives the shrieking saws
> Through smooth blocks of marble, hears from either bank a ceaseless din[10]

Archeologists authenticated the existence of this fourth-century sawmill for cutting marble and limestone on the Ruwar, a tributary of the Mosella River. Since Roman technology used neither cams and levers or cranks and converters, questions remained about the practice for cutting stone. The Roman historian Pliny provided a clue for cutting marble. Begin with a wire, the ends of which are pulled together. Use sand as an abrasive, thus forming a bandsaw. It becomes operational by looping the wire over two pulleys, one of which is geared to the power of the waterwheel.

> The power applied to the wire by the driving pulley should be great enough for it to abrade its way through the stone, and the particles of sand should provide enough friction to prevent its slipping on the pulleys. Raising the block on which the stone is placed, by means of appropriate gearing and at a rate matching the cutting rate of the wire, produces a fully automatic sawmill.[11]

During the latter decades of the twentieth century, archeologists discovered two triple-helix turbines for powering waterwheels in Roman Tunisia. For both turbines to operate, "the waterwheel was placed in the bottom of a water-filled, cylindrical pit entered tangentially by the inflow channel. The rotating water column makes this mill a true turbine."[12] Historians thought that inventors built the first turbines during the sixteenth century. The discovery of their existence in Roman Tunisia provided additional evidence about the high level of technological development during antiquity.

The Roman use of waterpower in hydraulic mining broke up alluvial deposits to uncover silver and gold. Water-powered stamp mills crushed ore in its Iberian Peninsula operations. " Roman mining saw some of the most advanced and large scale applications of technology to economically critical work ever to be practiced before the European industrial revolution, and some of the most impressive investment in infrastructural engineering works."[13]

As historian Andrew Wilson has pointed out, ancient technology made significant contributions to economic development. Controlling the power generated by moving water led to geared water-lifting machinery beginning in the third century BCE. Using animal power for lifting to irrigate the land was followed by water-powered wheels that automated the process. Milling flour using the power of waterwheels and trip hammers for crushing ore became commonplace by 100 CE. Water-powered saws became available by 300 CE.

Sixth- and seventh-century charters from Merovingian Europe (457–751 CE) suggest that watermills did not suddenly appear during this era. Documents from the Roman Empire prove that their use appeared centuries before. Clusters of 200 to 400 people with an available supply of running water warranted the construction of a mill to lessen the burden of human and animal labor. As more people congregated in towns and cities, greater numbers of watermills supplied residents with ground flour.[14] "The diversified application of water-power was evidently not originally a medieval phenomenon, and many of the machines once thought part of the 'medieval industrial revolution' were to be found in the Roman world."[15]

The ancient world left an impressive list of technological innovations. They included water-powered chains of pots for lifting water, the dough-kneading machine, olive crushing and olive pressing mills, the vertical and horizontal watermills, hydraulic pumps and the use of waterpower to mass-produce tiles, concrete, pottery and of course, bread.[16]

Water technology in ancient China

The Chinese used industrial vertical waterwheels as early as the 100 CE, long before they learned of developments in the West. Trip hammers and bellows dated from 31 CE were recorded in the *Hou Han Shu* and described by historian of China Joseph Needham. Chinese farmers used horizontal and vertical wheels powered by water to grind rice, wheat, beans and seeds. Its newly developed metallurgical industry used water-powered bellows. The establishment of the Silk Road connecting East and West allowed for the transfer of technology. "It seems likely that the compartmentalized waterwheel must have also been known in China before the second century, the knowledge of which most likely arrived there via the recently established Silk Road anything up to three centuries previously."[17]

Waterpower in Europe's early medieval economy

Attacks from tribes on its eastern boundaries and internal conflict led to the fall of the Roman Empire. As a result, much of the knowledge and application of milling technology and the Roman social structure that supported it disappeared. The disintegration of infrastructure, including transport routes, a uniform Latin language, and common administrative and legal codes that bound the Empire together fragmented. Clearly, some mechanical knowledge survived in Latin texts, and possibly more was passed on from one generation to the next by verbal communication. "That some of those technical advances made by the Romans either remained unknown to medieval Europeans (in the case of the turbine) or took as long as a millennium to be reintroduced (in the case of the cam and trip hammer)"[18] are examples that question our ideas about uninterrupted technological progress.

By 1000 CE in Europe, watermills held a ubiquitous place in the environment. Every village had its own watermill, averaging one mill for every 250 people.[19] Not until the 1400s and 1500s CE, however, did milling technology in the woolen cloth and ironworks industries become commonplace throughout medieval Europe. Before that time, only specific regions of the continent with enough financial support, flowing water and transport benefited from technological advances. Water-powered milling technology was in widespread use in the Roman provinces of Italia, Gallia, Hispania and Britannia centuries before it was reintroduced in medieval Europe. Long periods of fragmented and decentralized authority followed the empire's disintegration.

So, the reintroduction during the medieval period of many Roman innovations became sporadic and local in origin. In fact, cheaper, more equalitarian forms of power generation in the form of horizontal-wheeled watermills became dominant in the few centuries after the collapse of Rome. Landed aristocrats, as well as craftspeople, merchants and others entered milling occupations as workers and investors. The replacement of vertical-wheeled watermills with the cheaper to build and maintain horizontal structures appears in early medieval Italy, Spain, France and Britain. Some evidence exists regarding the existence of horse mills during this early period as well. Of the 6082 watermills identified in England's Domesday Book of 1066 CE, the majority was probably of the horizontal kind.[20]

At the end of the early medieval period, power became consolidated in the hands of aristocrats with large landholdings and monasteries. As the baron, Walter Fitz Robert (ca. 1126–1198 CE) from the baronial house of Clares ordered: "Erect no windmills or watermills to the detriment of existing mills."[21] Concentrated wealth and power dominated landholdings, preventing peasants and townspeople from participation in milling industries.

Profitable vertical watermills conveyed status. Expensive to build and maintain vertical watermills replaced horizontal versions favored by commoners. However, horizontal mills remained a highly adaptive power source throughout the Mediterranean region with links to the turbine. Where concentrated power was weak, these mills continued in use to the present day. As historian Richard Holt pointed out, however, the Black Death (1348–1349 CE) decimated England's working population. Labor costs rose, grain production declined and many millers abandoned their waterworks or sold off its movable parts. By the 1400s, investing in horse mills became a safer course to follow in the industrial sector. Animal power, an organic form of energy, continued to coexist with the more advanced power generated by water.

Regarding water-powered industrial mills, historian Adam Lucas has noted that from 770 to 1600 CE, "no more than four hundred industrial mills [have been] authenticated."[22] France and England possessed more than 80 percent of the total. By adding the German and Italian States, the percentage rises to 94 percent. However, the numbers will probably increase as archeologists and historians continue to locate additional sites in these four countries and in countries not accounted for yet. Of those for which we have data, 80 percent were mills for fulling cloth (i.e., the cleaning of cloth of its dirt, oils and any other contaminating substances) and forge mills. Sawmills, tool sharpening and tanning mills (i.e., curing leather) made up 12 percent of the total.

Adam Lucas described the process of fulling in the following way:

> Prior to the introduction of mechanical fulling in early Islamic societies by the tenth century, and in Europe by the eleventh century, most fulling had been done by foot, by hand, or with clubs wielded manually, inside a large trough that contained the cloth, a sufficient quantity of water to fully immerse it, and some kind of cleansing agent. The fullers would stand in this trough semi-naked, and, if fulling by foot, support their weight on the rim of the trough as they trampled the cloth underfoot. This process was fully understood by the Romans. Despite the spread of mechanical fulling throughout Western Europe by the end of the thirteenth century, a variety of these more primitive methods continued to be used in many countries until the twentieth century.[23]

Pounding cloth using human energy ended with the invention of recumbent wooden trip hammers. Waterpower moved a set of cams mounted on a drum that rotated on the axle of the waterwheel. Hammers rose and fell alternately on the cloth lying in a trough. Either gearing and/or the number of cams attached to the rotating drum regulated the frequency of the pounding. These innovations aside, a fulling mill mimicked the conventional vertical-wheeled watermill.[24]

The region that would become France led in the application of waterpower to industry in medieval Europe. It was home to the earliest documented water-powered fulling mill about 1050 CE, tanning mill about 1125 CE and a tool sharpening mill about 1225 CE. With England and Sweden, it shared the earliest documented forge mill. In about 1325 CE, it possessed a sawmill. France may also have been first in adapting waterpower to minting coins and cutting metal. Specific provinces of France with ample sources of running water and developed river and road transport systems included Normandy, Picardy and Champagne in the north, Burgundy in the east, and Languedoc and Provence in the south. At different times during the centuries in question, English and French kings and the Holy Roman Emperor controlled these provinces.

For England the authenticated sites for water-powered mills remains small. "Only fifty-five of the documented 1,647 powered mills or less the 3.5% were industrial mills, and all were fulling mills."[25] Unlike mills in France that included a number of industrial processes beyond those mentioned earlier, including paper, hemp, sawing, malt and oil, 90 percent of English mills ground grain into flour, while fulling cloth took much of the remaining 10 percent. None of the evidence provided here suggests a medieval industrial revolution even if we include the application of waterpower in the northern region of Italy. These would include Piedmont, Pistoia and Firenze, where waterpower represented 25 percent of its industrial applications.

By 1540 CE, a date suggested for the late medieval period in Europe, watermills were a visible presence wherever moving water attracted a sizable number of people. As early as 1300 CE, estimates of their number in England reached as high as 10,000. Throughout the continent, larger numbers made them prominent features on the land, and their numbers increased in the centuries to follow. In England, with an estimated population of 7 to 8 million, annual per person grain consumption reached about 12 bushels. Its mills, including water, wind, hand (about 20 percent of the total) and animal mills, turned much of its grain into flour. Some was consumed without grinding. Oats, legumes and barley were boiled to become a stew or porridge.[26]

Historian John Langdon has argued that medieval Europe made important industrial contributions, releasing humans from much drudgery using water, wind and animal muscle power as alternative energy sources. Historian Marc Bloch had argued that lords, large landowners, exercised their "suit of mill" prerogatives by insisting that their tenants use their mills for grinding grain. However, free tenants and others patronized mills of their choosing. These divisions and tensions among lords, tenants, and the growing urban population would continue "leading up to the key 'rupture' of the eighteenth and nineteenth centuries, when these patterns began to break down and new ones took their place."[27]

Waterpower and the Industrial Revolution

In the last decades of the eighteenth century, but possibly earlier in some regions of Europe, finding new sites for water-powered wheels became progressively more challenging as the textile industry grew and invested in new technology. With most favorable mill sites along fast-moving rivers already in use, the competition for potentially less desirable sites became more intense and more costly. The alternative of investing in steam engines to recycle water to storage ponds doubled the cost of power. If converting to steam power proved to be prohibitive, what alternatives were available to proprietors of watermills?

Owners focused on technological improvements in energy conversion to meet the increasing demand for machine-made cotton cloth. John Smeaton, founder of England's Society of Civil Engineers, led the eventual conversion of water and wind power to steam. He spearheaded the improved industrial design of water wheels, now made of iron rather than their traditional wood-based construction. Additionally, their dimensions changed significantly as well as their gearing, transmission, speed and the construction of larger and more durable buildings to accommodate much larger and more powerful iron wheels.

Large industrial waterwheels were dramatic converters of energy. The new cotton mill of Jedediah Strutt in Belper (a town in the valley of the River Derwent, England) 1804 symbolized the transition to large industrial water wheels. His new cotton mill derived its energy from a single wheel 18 feet in diameter and 23 feet wide. The five floors of the mill received power to turn over 4,000 spindles, plus reeling, carding, doubling and twisting machines. Power exceeded 100 horsepower much greater than for the most efficient low-pressure steam engines. With a workforce numbering more than 1,000, this mill had reached the pinnacle of energy produced by waterpower. The invention of modern turbine technology and hydropower replaced massive waterwheels of this size a century later.[28]

Despite technological achievements, sites for water-powered wheels remained scarce and expensive. The cost of maintenance could become prohibitive as rushing water undermined wooden mill buildings and their wooden internal works. To prevent such occurrences, waterwalls to create channels for the flowing water were built of expensive timbers to prevent erosion. The same was true for the construction of wooden weirs across a stream to raise the water level in a millrace to power a wheel. Carpenters and millwrights became primary occupations.

Although no data exist regarding the depletion of woodlands to build and repair mills, owners faced increasing costs. If we accept the figure of 10,000 mills in 1300 CE, the cost of construction alone would be about 350,000 pounds.[29] Power demands in an increasingly industrialized world would outstrip the capacities of immobile, fixed sites. Steam power would eventually break through these geological constraints. Environmental constraints, especially the depletion of woodlands, slowed the energy transition. At the same time, the growing number of mills needing energy quickened the transition.

Waterpower remained the main source of energy during the initial stages of the Industrial Revolution. Watermills and horse-powered mills shared equally in cotton manufacturing. In Oldham, England, six mills were built between 1776 and 1778, three using horses for power and three using water.[30] Waterwheels moved camshafts, crankshafts and pushrods, using their power for grinding grain, bark, pounding rocks, pumping water, boring holes, planning, cutting, sawing, fulling and dyeing textiles. In addition, they inflated bellows for iron making. With a series of new inventions, the spinning jenny (1770), Richard Arkwright's water frames (1771) and Edmund Cartwright's use of steam engines to drive textile looms in 1787, waterwheels continued to power England's textile industry until the middle of the nineteenth century.[31]

In metallurgy, waterpower became the primary energy source. Water-driven wheels occasionally provided the power for machines to drain the deep shafts of gold and silver mines. More often, such draining was accomplished by digging tunnels, horizontal to flooded mines, and allowing gravity to do its work. For copper, lead, tin and iron mines, buckets raised by human and horses or windlasses drained the mines. In smelting and fabricating metal from the thirteenth century onward, waterwheels provided the power for bellows whose blast of air into furnaces removed the impurities from the ore. In the case of pig iron and copper, the refining process required reheating and hammering to produce a workable product. Power to the bellows of the blast furnace and the power hammers of the refinery came from the waterwheels.

Historian Ian Blanchard described a highly productive hydraulic and watermill complex in Czarist Russia in the 1780s CE. In the government-owned Ekaterinburg iron and copper facilities in the Ural Mountains, a complex energy system of waterwheels powered all work. Sluices regulated and controlled the flow of the dammed-up waters of the Puitma River. By adjusting the amount of water passing through a central escape conduit, it altered water pressure flowing through two service channels. They drove a complex of waterwheels that powered the bellows of the blast furnaces and the power hammers of the refinery. They also provided the power for rolling mills and machine tools for cutting and drilling sheet metal. Forges, foundries and steelworks located on the same channels received their power from this same hydraulic energy source.[32]

The transition to steam and the early uses of coal in Europe

Russian ironworks and English cornmills used the enhanced technology of waterpower as the major energy supply. The hydraulic systems of water wheels, drive shafts and belts delivered large amounts of energy sufficient to power many of the mills and "factories" of the pre-factory age.[33] According to Ian Blanchard, the new Thomas Newcomen and James Watt steam engines failed to penetrate the industrial sector in the years prior to 1780 CE. In coal mining, steam engines excelled in draining deep pits, work formerly done by horse or man-powered windlasses, not by windmills. Also, the location of coal mines near ironworks improved the economic benefits of using steam engines in the smelting process. Improved roads, canals and steam-powered railcars, along with the supply network of sailing ships powered by wind, encouraged the use of Newcastle coal. As coal prices declined, price, not the superiority of steam over waterpower, drove technological progress. The flow of many energy sources contributed to the transition from manufacturing to industrialization.

Technological advances accompanied both forms of power generation. Substituting steam for water as an energy supply occurred slowly, but the transition to steam power became evident in the period after 1790 CE. However, beginning in the 1840s, water-driven turbines replaced waterwheels. By the 1890s, turbine-driven generators provided electricity. Innovations in hydraulic technology as a source of energy continued as a widening and deepening deployment of steam engines became evident.[34]

In Great Britain, according to historian Louis C. Hunter, the role of steam power in draining mines existed for 100 years before its industrial applications. Its numerous rivers provided sites for thousands of mills for grinding bread grains, but few possessed the rapid flow to accommodate manufacturing enterprises. Windmills dominated most high ground to grind corn. Despite the limitations imposed by geography and topography, and excluding such vital industrial activities as textiles, mining, ironworks and railways, all other factories possessed limited steam-power capacities in 1850.

In the 80 years that followed, however, the transition to steam power accelerated rapidly. Steam-powered machinery in the cotton textile industry used low-priced coal. In response, the number of mines supplying coal expanded quickly. Dye and bleach workshops depended on coal-fired vats. Glassworks, paper factories, salt and sugar refineries heated their boilers and vats using locally marketed coal by 1870 CE. In 1930, at the outset of the global Great Depression, waterpower represented but 1.4 percent of the country's total industrial and central-station power.[35] In the early stages of factory development, however, waterpower became an essential source of energy.

In France, an abundance of timber, a shortage of coal and a network of navigable and fast-flowing rivers distinguished its factory development from that of Britain. The scarcity of domestic coal and the high cost of imported coal made the use of James Watt's inefficient steam engine prohibitive, favoring Arthur Woolf's compound engine for its fuel-saving quality.

A decade later, when Britain and the United States increased their steam power capacity, France's steam power accounted for less than one-third of its energy use. Much more effort was placed on the technological improvements of its water wheels. Inventors Jean-Victor Poncelet and Benoit Fourneyron perfected the undershot wheel with curved floats and the water turbine, respectively. A comprehensive industrial survey conducted from 1861 to 1865 concluded that waterpower exceeded steam power by a ratio of two to one. In the next four decades, however, France's transition to steam power occurred more rapidly than it did in the United States.[36]

Historian Hunter noted, "There is no evidence that steam power is intrinsically superior to waterpower in its operation of driven machinery."[37] Yet, developments pointed in the opposite direction. The superiority of the steam engine in its stationary form and its mobile applications as railway locomotives and steam-powered ships revolutionized industrial production in the nineteenth century. Advances in the application of wind power to marine transportation during the era of "clipper ships" (1849–1863) delayed the transition to steam vessels. However, their superiority in river traffic, replacing barges and flatboats and on the high seas with coaling stations located strategically around the globe, greatly enhanced the globalization of capitalism. Steamships linked markets of finished goods with suppliers of raw materials.

Watermills in the United States

The energy gained from falling water powered nineteenth-century industrialization in the United States. Factories grew up along major Eastern cities where the falling water of its rivers attracted capital investment in new mills. New England led the way, according to the U.S. Census, in 1880, claiming that the region possessed "at least ten developed powers of 10,000 theoretical horsepower or over during working hours . . . at least eighteen of over 3,000 horsepower continuously, and at about twenty of over 2,000 horsepower continuously."[38] With the publication of the first waterpower volume of the Tenth U.S. Census in 1885, its introduction made the following claim:

> It is probably safe to say that in no other country in the world is an equal amount of water-power utilized, and that, not only in regard to the aggregate employed, but in regard also to the number and importance of its large improved powers, this country stands pre-eminent.[39]

As historian of technology Patrick M. Malone has pointed out, the mills at Lowell, Massachusetts, became one of the country's preeminent centers for textile and machinery manufacturing. New England mills at Manchester and Nashua, New Hampshire; Lawrence, Massachusetts; Biddeford and Saco, Maine; and Chicopee and Holyoke, Massachusetts, followed the industrial model of Lowell. Beginning in 1821, investors from the Boston Manufacturing Company, who owned and successfully operated mills on the Charles River in Waltham, Massachusetts, saw the potential for building a much larger industrial complex at the Pawtucket Falls on the Merrimack River.

As early as 1825, the founding directors of the Merrimack Manufacturing Company saw the potential energy of the river. So, to maximize the return on its investments, it created the Proprietors of Locks and Canals to sell land and to lease waterpower to new firms along the Merrimack. In effect, Locks and Canals rented a specific amount of water during a working day and in return built and maintained the canals. New factory owners customarily employed Locks and Canals to construct worker housing, and to

> build and equip their mills, they usually chose to let the canal company build their head gates, trash racks, races, wheel pits, waterwheels, and power transmission systems as well. A steady progression of building projects and lucrative machinery orders made *Locks and Canals* a highly profitable company.[40]

Expanding on the successful prototype in Waltham, Boston investors took on an industrial venture in size, scale and power previously unknown in the United States or anywhere else in the world.

> In place of the traditional millstream, often no more than a brook or a creek, (capable of harnessing no more than 100 horsepower in a single installation), rivers hundreds of feet wide, with volumes of flow measured in tens of thousands of cubic feet per second, (cfs) and (measured in thousands of horsepower) had to be harnessed to the tasks of industrial production within great mill structures.[41]

In order to bring the seasonally swollen Merrimack River under control, dams, guard gates and millraces larger than hydraulic canals had to control the volume of water and direct it to the penstocks and waterwheels of the textile mill. With its energy spent, the water returned to the riverbed below the factory. By continuously improving and expanding on a previously underutilized canal, by 1848, 10 financially successful corporations had built a 5-mile-long, two-level canal system and harnessed the energy of the river. In all, 10 power distribution canals with a range of 500 to 6,500 feet long, combined to reach a total of 5 miles. An 11th corporation focused on building machinery to service the textile mills downriver.

These wealthy Boston merchants made these capital investments as early cotton industrialists. In the United States of the nineteenth century, the cotton-planter class possessed unequaled political power in Congress and access to large tracks of land and slave labor. By 1830, 1 million people, or one in every thirteen people in the United States, were slaves, planting and picking cotton in the southern states. Initially, the Lowell mills produced course cotton to clothe slaves, replacing cloth previously purchased by slave owners from India. "'Lowell' became the generic term slaves used to describe course cottons."[42]

Although the development and industrial capacity of the Lowell mills will remain the primary focus of this section, as cotton became a global commodity, mills in the United States opened in Rhode Island (1790), New Jersey (1791), Delaware (1795), New Hampshire (1803), New York (1803), Connecticut (1804) and Maryland (1810). The U.S. Census listed 269 cotton mills with a total of 87,000 spindles in 1810. By 1860, cotton mills represented the country's most important industry in terms of capital invested, workers employed and the net value of its cloth.[43]

The forced labor of African slaves and their offspring planted and picked cotton, in most cases under the most oppressive conditions in which failure to perform or disobedience was met with brutal assaults or death. Having one's clothing stripped off and being staked to the ground for a lashing with a whip was a common form of punishment. At the other end of the production process were the workers in the mills. Who were they? According to historian Sven Beckert, mill workers employed in Lowell, Manchester (England), or (New Hampshire), Barcelona (Spain) or Chemnitz (Germany) worked under similar conditions and experienced degrading forms of punishment:

> The world had seen extreme poverty and labor exploitation for centuries, but it had never seen a sea of humanity organizing every aspect of their lives around the rhythms of machine production. For at least twelve hours a day, six days a week, women, children, and men fed machines, operated machines, repaired machines, and supervised machines. They opened tightly packed bales of raw cotton, fed piles of cotton into carding machines, they moved the huge carriages of mules back and forth, they tied together broken yarn ends, they removed yarn from filled spindles, they supplied necessary roving to the spinning machines, or they simply carried cotton through the factory . . . In one English mill, of the 780 apprentices (mostly young girls and boys) recruited in the two decades after 1786, 119 ran away, 65 died, and another 96 had to return to overseers or parents who had originally lent them out.[44]

By this time, Lowell's industrial capacity exceeded that of any water-powered cotton mills in the country. To calculate the energy needed to operate such a large complex of factories, the term *mill power* replaced the conventional *horsepower* that originated with James Watt to indicate the amount of work that would otherwise be required of 8 to 10 adults. Mill power was based on the power required to drive 3,584 spindles and the looms to weave the spun yarn. Translated into waterpower terms, the so-called Lowell standard described mill power as 25 cubic feet of water per second falling thirty feet and equaling 62.5 net or

85.2 gross horsepower. By 1862, massive waterwheels powered 380,000 spindles and employed 9,000 girls and women and 4,000 boys and men. With preparatory and weaving machinery together with the network of hydraulic canals along the 5-mile river corridor, they produced "a variety of textile products, including enough cotton cloth every day to 'extend in a line about a yard wide' for two hundred miles."[45]

To achieve this level of productivity, however, required a major capital investment and the energy of human labor, which was exhausting and dangerous. Men using shovels, hammers and drills dredged the bottoms of the Pawtucket and Merrimack canals, while other men with wheelbarrows and the assistance of horse- and oxen-drawn wagons hauled out tons of sediment and rock. About 3 tons of gunpowder blasted away ledge to widen the enhanced canal system. There, stonemasons built retaining walls. When the canals entered swampland, dredging, excavation and masonry created the width and depth required for the two-tiered canal system. Injuries, some requiring amputations, were a constant hazard from exploding gunpowder to change the geology of the canal system. With the building of the Northern Canal and the Pawtucket Dam in 1846–1847, "seven steam engines came into play for pumping water, dredging, and other jobs."[46]

The transition from Breastshot waterwheels to outward flow turbines

As Patrick Malone has pointed out, as flow on the Merrimack River declined during dry seasons, mill operators installed steam boilers and engines to maintain peak production. The installation of steam-powered machinery as a supplement to waterpower took place as mill owners converted their mills from breast wheel to higher efficiency turbines. By 1860, only 3 of the 10 textile corporations used breast wheels.[47] As an aside, engineers distinguish breast wheels from vertical and horizontal waterwheels in the following way. In breast wheels, power is generated by water hitting the back of the wheel. Vertical waterwheels can be immersed in water and powered by the force of the flow (undershot) or (overshot) in which water is conducted through a wooden chute above the wheel. In the horizontal wheel, water enters through a deep chute and cascades onto the wheel's oblique paddles.[48]

In addition to the transition to more efficient turbines, mill owners had access to surplus water, available when they used up their permanent "mill powers." Ironic as it may seem with efficiencies and surplus, steam power became a necessary addition to the power mix. As James B. Francis, chief engineer for Locks and Canals, pointed out in 1854, a mill was "to be driven by water whenever we have it to spare for that purpose, the remainder of the year say six months to be driven by steam."[49]

Since steam engines required burning coal to turn water into steam, air pollution in Lowell would become a by-product of industrialization. The number of steam engines increased from 29 in 1866 to 88 in 1881. However, most were kept in reserve with the availability of inexpensive surplus water. The availability of both coal-fired steam and surplus water allowed for the continuous flow of production from raw cotton to finished fabric. A reduction in the production process caused by an unpredictable loss of waterpower would reverberate throughout the factory, causing slowdowns, work stoppages, unemployment and the failure to fill work orders for fabric. Running at full capacity was essential for maximum productivity.[50]

The French engineer Benoit Fourneyron designed a functional water turbine in 1832. Uriah Boyden, using this original design, developed an outward flow turbine for the Appleton mills in Lowell in 1844. As described by Patrick Malone, pressurized water

> entered the center of the wheel from above, through an iron conduit called a penstock, and was directed outward by fixed curved guides. Guides surrounded the critical moving part, a runner with curved buckets, or vanes. Because the runner reacted to the pressure of the water on its buckets, this class of wheel was called a *reaction turbine*.[51]

At one of the Appleton Company mills in 1844, a new Boyden turbine operated with a tested efficiency of 78 percent and generated 75 horsepower. By operating underwater, they captured the entire power of the fall. Their metal construction, a smaller size, eliminated the large wheel pits and eliminated the dangers associated with unguarded breast wheels, some of which reached 80 feet high. Operating at high speeds, turbines eliminated the costs of gears and pulleys needed to change the speed of breast wheels. They represented a technological innovation that would ultimately replace overshot and breast wheels.

It took only four years after the first installation at Appleton Company for chief engineer Francis to comment "that no new breast wheels be put in the mills."[52] In fact, mill operators replaced breast wheels with Boyden turbines when they malfunctioned due to use or age. Or they replaced them with new industrial capacity to meet the growing demand for textiles. One mill corporation continued to use one of the city last large breast wheels into the 1890s. However, the transition to turbines gained momentum throughout the mid-nineteenth century. In 1858, 56 Boyden turbines generating 12,000 horsepower and turned thousands of spindles in the Lowell mills. Within a few years, textile mills in Manchester, Lawrence and other industrial cities used turbines locally manufactured at Chelmsford's Gay, Silver and Company machine shop.

The primacy of waterpower in the development of industrialization in the years prior to the Civil War in the United States was reflected in the value of its tenfold increase in manufacturing output while the population rose fourfold. While Lowell and its counterparts in other eastern cities dominated discussions of large-scale mechanized production, the country's village and town histories began with the construction of community-based watermills.

According to historian Merle Curti, "[a]lmost all the villages of the county were plotted at points deemed good for mill-sites."[53] Abundant and cheap, water's irregularity and unreliability, given the seasonal changes in flow, suited an economy in which riverboats and canal barges faced similar limiting conditions. Steam-powered ships, the growing network of railroads and the concentration of large-scale industries in cities changed the power mix. Waterwheels and turbines were replaced by coal-fired steam engines.

During that decade, machine shops in most states manufactured turbines, with 20 shops entering Philadelphia's Fairmont Waterworks competition of 1859–1860. In Massachusetts in 1875, they powered 82 percent of all waterwheels. By 1878, 80 manufacturers fabricated hydraulic turbines in the United States.[54] However, with much of the country's growing industrial capacity taking place in cities without appreciable waterfronts, coal-fired steam engines began to catch up with hydraulic turbines. By 1879, steam-powered turbines represented 64 percent of all installed horsepower. The transition was slowed by newly designed wheels with superior efficiencies as stronger metals continued to replace worn-out equipment. Wooden waterwheels lasted a decade at most while metal turbines continued to operate on average for 30 years before replacement.

Fifteen years after the construction of the Lowell mills, more than 100 steam cotton mills became operational in the United States, a few of which matched the Merrimack River complex in size and output.[55] Coal-fired steam boilers and machines continued to replace water-powered machinery in the generation after the American Civil War (1861–1865). Steam power represented 96 percent of the growth in industrial capacity, from 2.4 million to 6 million horsepower from 1869 to 1889. During this boom period in the country's Industrial Revolution, waterpower grew by a paltry 3 percent.[56] By 1900, however, waterpower received a boost from the emergence of hydroelectricity, the transmission of electrical power to factories, commercial buildings and urban households. With advances in hydroelectric technology after 1900, single wheels generated the capacity of the entire direct – drive waterpower system of the Lowell, Manchester or Holyoke mills.

Location, regarded as the proximity to rivers capable of generating lots of waterpower, was of primary importance to capitalists searching for the most viable place to construct a mill. With the urbanization of manufacturing in the years following the Civil War, the economy experienced a widespread shift to steam power enhanced by the boom in railroad construction, urban population growth and the growth of large size industrial operations. During a 50-year period, from 1849 to 1899, the country's population tripled, but its urban population literally expanded eightfold. Due partly in response to the thirteenfold increase in urban manufacturing, it helped to explain the tenfold increase in horsepower, caused mostly by burning coal in steam boilers to power machines. To accommodate the population boom, agricultural productivity grew fourfold.[57]

Waterpower remained stable but stagnant as factories made the nearly complete transition to steam power in 1880. Fall River's textile plants eclipsed Lowell's in the 1870s, with their conversion to steam, powering twice the number spindles as those at the Lowell works. All ironworks, coal blast furnaces, including the new Bessemer and open-hearth steelworks of the Ohio Valley, employed steam power. Waterpower, no matter the size of the wheels or turbines, could not meet the power requirements of large-capacity blast furnaces and massive rolling steel mills. The transition was not restricted to the iron- and steelworks, however, as textiles, grist, and lumber industries made similar moves.

Yet, despite the rate of replacement, especially in these large-scale industries, and the stagnation in the expansion of water horsepower in the 1880s, from 1869 to 1909 water horsepower increased by 61 percent nationally and 109 percent in New England's textile industries where the cost of power remained a meaningful consideration.[58] With each passing decade, however, the cost differential between these two forms of mechanical energy narrowed because of the efficiencies gained by technological innovations in large-scale steam engines. As Charles Main, past president of the American Society of Mechanical Engineers and at the time superintendent of the Lower Pacific Mills in Lawrence, Massachusetts, concluded in 1890

> That the balance of advantages and cost combined is in favor of steam power for textile manufactures is proven at the present time by the erection of steam mills almost entirely, while there are still undeveloped water powers which are available.[59]

Other important considerations accelerated the transition to steam power, not the least of which was the increasing demand from growing cities for clean potable drinking water. Damming rivers and creating reservoirs and ponds to ensure a steady flow of water to waterwheels changed their ecology. Spawning fish needed access to upstream pools, and juveniles needed to gain access to downstream food sources. Dams restricted their movements and that of the river, exposing it to concentrated sunlight and raising the river's temperature. Once used as "sinks" for sewage and factory refuse, flowing river water, according to sanitation engineers, would provide an adequate supply of safe drinking water for growing, dense urban populations.

In addition to the quest for cleaner drinking water, other interests for water use competed with mill owners. The accelerated clearing of land for farming raised land and crop values and pitted agricultural interests against mill flowage into bottomlands and meadows. Loggers, lumbermen, sawmill owners, seasonal fishermen, barge and boatmen also competed with mill owners for water rights. Statutory law and judicial decisions challenged the rights of streamflow, oftentimes limiting usage by mill operators.

Conclusion

As noted throughout these initial chapters, from antiquity until the last decades of the nineteenth century, human, animal, wood, wind and water (hydraulic) sources of energy changed little until the invention and application of the steam engine to do work. Turning water into steam required combustible material. In the Americas, woodlands provided firewood for this purpose. In Europe, fossil coal replaced wood sources much earlier, due to extraction and transportation costs. Despite the smoke, grime and coal ash, by-products of combustion, mill operators were free for a time to contaminate the atmosphere and the surrounding land with impunity.

Despite complaints from local groups, smoke control ordinances to protect inhabitants from such "nuisances" would wait until the costs, both environmental and political, outweighed the benefits, sometime in the early decades of the twentieth century. The transition from a millennia-long organic energy regime to one based on fossil coal would be gradual, as we will see in the chapter to follow, but it would be transformative.

At the Lowell millworks, the transitions from waterwheels to turbines to steam power overlapped mostly. Locks and Canals continued to maintain and operate the canals to supply water to the mill complexes and by 1918 installed new hydroelectric generators to the turbines. Management delayed plans to build a central electric power station near the Pawtucket Falls in 1894. Declining revenues, the Great Depression (1929–1938) resulted in the closure of all but three mill complexes. By the mid-1950s, they closed as well. Yet, hydroelectric power remained a salable commodity into the distant future as Chapter 7 will make clear.[60]

Notes

1 John Langdon, *Mills in the Medieval Economy: England 1300–1540* (New York: Oxford University Press, 2004), 13–15, 19, 24, 34–37.

2 Adam Lucas, *Wind, Water, Work: Ancient and Medieval Milling Technology* (Leiden: Brill, 2006), 1.

3 Lewis Mumford, *Technics and Civilization* (New York: Harcourt, Brace & Co., 1934), 112, 113–118. Marc Bloch, "The Advent and Triumph of the Watermill," *Land and Work in Mediaeval Europe: Selected papers by Marc Bloch* (Berkeley and Los Angeles: University of California Press, 1967), 136–168.

4 The quote is taken from Lucas, *Wind, Water, Work*, 44. Who cites the English translation of Bloch's paper by J.E. Anderson, in Bloch (1967), 146.

5 Orjan Wikander, "The Watermill," in *Handbook of Ancient Water Technology*, ed. Orjan Wikander (Leiden, 2000), 371–400. Adam Robert Lucas, "Industrial Milling in the Ancient and Medieval Worlds: A Survey of the Evidence for an Industrial Revolution in Medieval Europe," *Technology and Culture* 46, no. 1 (January 2005), 1–30. Andrew Wilson, "Machines, Power and the Ancient Economy," *The Journal of Roman Studies* 92 (2002): 1–32.

6 Andrew Wilson, "Machines, Power and the Ancient Technology," 10.

7 Stuart Fleming, "Gallic Waterpower: The Mills of Barbegal," *Archaeology* 36, no. 6 (November/December 1983): 68.

8 Orjan Wikander, "Where of Old all the Mills of the City Have Been Constructed: The Capacity of the Janiculum Mills in Rome," in *Ancient History Matters*, eds. Karen Ascani, Vincent Gabrielsen, Kirsten Kvist (Rome: L'erma di Bretschneider, 2002), 130.

9 R.H. Sellin, "The Large Roman Water-Mill at Barbegal (France)," *History of Technology* 8 (1983): 100–101.

10 *Ausonius*, trans. H.G. Evelyn White, 2 vols. (London, 1919), 1.253, bk. 10, *Mosella*, lines 361–364. For the dating see l: Xvii, 1:223. Quoted in D.L. Simms, "Water-Driven Saws, Ausonius, and the Authenticity of the Mosella," *Technology and Culture* 24, no. 4 (October 1983): 635.

11 D.L. Sims, "Water-Driven Saws in Late Antiquity," *Technology and Culture* 26, no. 2 (April 1985), 275.
12 Orjan Wikander, *Handbook of Ancient Water Technology* (Leiden: Brill, 2000), 377.
13 Wilson, "Machines, Power and the Ancient Economy," 17.
14 Orjan Wikander, "Sources of Energy and Exploitation of Power," in *The Oxford Handbook of Engineering and Technology in the Classical World*, ed. John Peter Oleson (New York: Oxford University Press, 2008), 149.
15 Ibid., 31.
16 Lucas, *Wind, Water, Work*, 45.
17 Ibid., 58.
18 Ibid., 321.
19 Paolo Malanima, "Energy Systems in Agrarian Societies: The European Deviation," in *Economia e Energia*, 83.
20 Richard Holt, *The Mills of Medieval England* (Oxford: Basil Blackwell, 1988), 112–113.
21 Edward J. Kealey, *Harvesting the Air: Windmill Pioneers in Twelfth-Century England* (Berkeley: University of California Press, 1987), 109.
22 Lucas, *Wind, Water and Work*, 213.
23 Ibid., 244.
24 Ibid.
25 Ibid., 223.
26 Langdon, *Mills in the Medieval Economy*, 145.
27 Ibid., 305.
28 Peter Mathias, "Economic Expansion, Energy Resources and Technical Change in the Eighteenth Century: A New Dynamic in Britain," in *Economia e Energia*, 25.
29 Langdon, *Mills in the Medieval Economy*, 179.
30 Rene Leboutte, "Intensive Energy Use in Early Modern Manufacture," in *Economia e Energia*, 560.
31 Ibid., 561–562.
32 Ian Blanchard, "Water-and Steam Power: Complementary and Competitive Sources of Energy," in *Economia e Energia*, 726.
33 Ibid., 727.
34 Ibid., 731.
35 Ibid., 164.
36 Ibid., 188–189.
37 Louis C. Hunter, "Waterpower in the Century of the Steam Engine," in *America's Wooden Age: Aspects of its Early Technology*, ed. Brooke Hindle (Tarrytown, NY: Sleepy Hollow Restorations, 1975), 163.
38 As quoted in Louis C. Hunter, *A History of Industrial Power in the United States, 1780–1930, Volume One: Waterpower in the Century of the Steam Engine* (Charlottesville: University of Virginia Press, 1979), 205. (U.S. Census Office, Tenth Census, 1880, vols. 16–17, *Reports on the Water-Power of the United States* (Washington, D.C., 1885,1887), pt.1, xxv–xxxvii).
39 As quoted in Patrick M. Malone, *Waterpower in Lowell: Engineering and Industry in Nineteenth-Century America* (Baltimore, MD: The Johns Hopkins University Press, 2009), 1–2.
40 Ibid., 42.
41 Hunter, *A History of Industrial Power in the United States*, 206.
42 Sven Beckert, *Empire of Cotton: A Global History* (New York: Alfred A. Knopf, 2015), 147.
43 Ibid., 140.
44 Ibid., 178.
45 Malone, *Waterpower in Lowell*, 2.
46 Ibid., 92.

47 Patrick M. Malone, "Surplus Water, Hybrid Power Systems, and Industrial Expansion in Lowell," *IA The Journal of the Society for Industrial Archeology* 31, no. 1 (2005): 31.
48 Orjan Wikander, "The Water Mill," in *Handbook of Ancient Water Technology*, vol. 2, Orjan ed. Orjan Wikander (Leiden: Brill, 2000), 274–276.
49 Malone, "Surplus Water, Hybrid Power Systems, and Industrial Expansion in Lowell," 32.
50 Ibid., 32–33.
51 Malone, *Waterpower in Lowell*, 107.
52 Hunter, *A History of Industrial Power in the United States*, 331.
53 As quoted in Hunter, "Waterpower in the Century of the Steam Engine," 179.
54 Ibid., 345.
55 Ibid., 540.
56 Ibid., 481.
57 Ibid., 489–490.
58 Hunter, "Waterpower in the Century of the Steam Engine," Volume One, in *A History of Industrial Power in the United States*, 522.
59 Ibid., 528.
60 Malone, *Waterpower in Lowell*, 221.

References

Beckert, Sven. *Empire of Cotton: A Global History*. New York: Alfred A. Knopf, 2015.

Blanchard, Ian. "Water-and Steam Power: Complementary and Competitive Sources of Energy." In *Economia e Energia, SECC.XIII-XVIII, Instituto Internazionale Di Storia Economica*, edited by Simonetta Cavaciocchi, 725–735. Firenze, 2002.

Bloch, Marc. "The Advent and Triumph of the Watermill." In *Land and Work in Mediaeval Europe: Selected Papers by Marc Bloch*, 136–168. Berkeley and Los Angeles: University of California Press, 1967.

Fleming, Stuart. "Gallic Waterpower: The Mills of Barbegal." *Archaeology* 36, no. 6 (November/December 1983): 68–87.

Holt, Richard. *The Mills of Medieval England*. Oxford: Basil Blackwell, 1988.

Hunter, Louis C. "Waterpower in the Century of the Steam Engine." In *America's Wooden Age: Aspects of its Early Technology*, edited by Brooke Hindle, 161–205. Tarrytown, NY: Sleepy Hollow Restorations, 1975.

Hunter, Louis C. *A History of Industrial Power in the United States, 1780–1930, Volume One: Waterpower in the Century of the Steam Engine*. Charlottesville: University of Virginia Press, 1979.

Kealey, Edward J. *Harvesting the Air: Windmill Pioneers in Twelfth-Century England*. Berkeley: University of California Press, 1987.

Langdon, J. *Mills in The Medieval Economy: England 1300–1540*. New York: Oxford University Press, 2004.

Leboutte, R. "Intensive Energy Use in Early Modern Manufacture." In *Economia e Energia, SECC.XIII-XVIII, Instituto Internazionale Di Storia Economica*, edited by Simonetta Cavaciocchi, 547–575. Firenze, 2002.

Lucas, Adam. "Industrial Milling in the Ancient and Medieval Worlds: A Survey of the Evidence for an Industrial Revolution in Medieval Europe." *Technology and Culture* 46, no. 1 (January 2005): 1–30.

Lucas, Adam. *Wind, Water, Work: Ancient and Medieval Milling Technology*. Boston: Brill, 2011.

Malanima, Paolo. "Energy Systems in Agrarian Societies: The European Deviation." In *Economia e Energia, SECC.XIII-XVIII, Instituto Internazionale Di Storia Economica*, edited by Simonetta Cavaciocchi, 61–99. Firenze, 2002.

Malone, Patrick M. "Surplus Water, Hybrid Power Systems, and Industrial Expansion in Lowell." *IA: The Journal of the Society for Industrial Archeology* 31, no. 1 (2005): 23–40.

Malone, Patrick M. *Waterpower in Lowell: Engineering and Industry in Nineteenth-Century America*. Baltimore, MD: The Johns Hopkins University Press, 2009.

Mathias, Peter. "Economic Expansion, Energy Resources and Technical Change in the Eighteenth Century: A New Dynamic in Britain." In *Economia e Energia, SECC.XIII-XVIII, Instituto Internazionale Di Storia Economica*, edited by Simonetta Cavaciocchi. Firenze, 2002.

Mumford, Lewis. *Technics and Civilization*. New York: Harcourt, Brace & Co., 1934.

Sellin, R.H. "The Large Roman Watermill at Barbegal." *History of Technology* 8 (1983).

Sims, D.L. "Water-Driven Saws in Late Antiquity." *Technology and Culture* 26, no. 2 (April 1985): 275–276. doi: 10.2307/3104347.

Wikander, Orjan. *Handbook of Ancient Water Technology*. Leiden: Brill, 2000.

Wikander, Orjan. "The Watermill," In *Handbook of Ancient Water Technology*, edited by O. Wikander, 371–400. Leiden, 2000.

Wikander, Orjan. "Where of Old all the Mills of the City Have Been Constructed: The Capacity of the Janiculum Mills in Rome." In *Ancient History Matters*, edited by Karen Ascani, Vincent Gabrielsen, and Kirsten Kvist. Rome: L'Erma di Bretschneider, 2002.

Wikander, Orjan. "Sources of Energy and Exploitation of Power." In *The Oxford Handbook of Engineering and Technology in the Classical World*, edited by John Peter Oleson, 136–157. New York: Oxford University Press, 2008.

Wilson, Andrew. "Machines, Power and the Ancient Economy." *The Journal of Roman Studies* 92 (2002): 1–32. doi: 10.2307/3814857.

Part II

The mineral energy regime

3 The coal revolution

The transition from an organic to a mineral economy

Introduction

Coal's transportability and its availability signaled a shift from an organic economy of muscle, wood, wind and water to a mineral economy that would initially be dominated by coal production and consumption but eventually include petroleum and natural gas use. Coal is the product of solar energy formed by decaying giant tree ferns and other vegetation that flourished in the great hot delta swamps of the Paleozoic Era, some 300 million years ago. Over the many thousands of years, volcanic eruptions, uplift, subsidence and flooding covered the vegetation with thick layers upon layers of alluvial mud.

Under the weight of these sediments, the pressure and heat characteristic of all chemical processes over the millennia converted the vegetation into rock, shale and a variety of carbon-rich substances. By excluding air in the 300 million-year process, most of the plant carbon is retained, rather than being lost as carbon dioxide. *Peat*, that spongy, water-soaked substance contains the lowest carbon content of about 60 percent. *Lignite* contains less water and a bit more carbon content, sometimes referred to as brown coal. *Bituminous*, with a larger concentration of carbon at 80.5 percent and less hydrogen at 5.3 percent and 12 percent oxygen, is truly coal. *Anthracite*, with most of the hydrogen and oxygen squeezed out, is solidly carbon at 96 percent. Although difficult to ignite, it burns at much higher temperatures than the other three categories of coal and will become the focus of this chapter. Coal also contains trace amounts of nitrogen, sulfur and mineral ash. When burned, these elements enter the atmosphere as pollutants.[1]

Coal use in antiquity: China and Imperial Rome

Ancient China

A massive tectonic eruption in northern China about 300 million years ago ripped apart the plant-covered terrain and replaced it with yellow soils and deserts. Intense pressure from the shifting and accumulating terrain above cooked the plant life over millions of years producing vast reserves of brown lignite and black bituminous coal. Throughout much of the world where coal is mined, tectonic

Figure 3.1 Vietnamese street venders in the capital city of Hanoi molding coal dust and straw to make bricks of coal for heating and cooking.

Source: Photograph taken by the author in 2014.

events shaped the mineral composition of what would become North America, Europe's Britain and Germany, Australia and China. Coal fueled China's energy history and its economic resurgence during the past decades.

Based on limited evidence, Ice Age communities in Europe burned lignite 73,000 years ago. Bronze Age China burned coal for bronze casting. Burned coal, not charcoal, found at four sites between 2600 and 1500 BCE confirm this conclusion. Meeting its energy needs began with the opening of the Fushan mine in northeastern China 1000 BCE for cooking, heating and metallurgy. Carved coal ornaments became items for trade in regional marketplaces. The large contribution of charcoal to the expansion of bronze casting required the burning of the region's vast woodlands. Overconsumption negated ideas about the inexhaustibility of its forests and led to one of the world's first transitions from an organic energy regime to one based on mineral coal. Deforestation accelerated that transition long before the industrial age. During the Han dynasty (206 BCE to 200 CE) considered ancient China's golden age, coal became the fuel that smelted iron ore in the production of steel. From that early beginning, coal became the country's primary energy source. China's consumption of 2.81 billion metric tons of coal approached the level consumed by the rest of the world in 2017.

Figure 3.2 Molded brinks of coal for residential and commercial sale.

Source: Photograph taken by the author in 2014.

Currently, it supplies about 70 percent of the country's energy, an increase of 10 percent in the last decade.

Roman Britain

In a world of abundant land and sparse populations amounting to a few people for each square mile, trees provided all the timber that humans needed for heating and cooking. Wherever church clergy or lords of the manor restricted access to trees, branches from fallen trees served as a viable energy source for the landless peasants and freemen and -women. Because coal existed in outcroppings, its use remained limited. In South Wales during the Bronze Age (3700–500 BCE), evidence existed of its use in cremations. However, its applications proved to be more widespread in Roman Britain (43–410 CE). It heated the military garrisons and forts on the northern frontier bordering Scotland, at the Antonine Wall and Hadrian Wall (122 CE). Aside from its military use, ample evidence exists of its civilian use for heating and cooking as well, across the north, southwest and southeastern parts of Roman Britain.

Metalsmiths clearly experienced the benefits of coal gathered from nearby outcrops compared to charcoal in working molten iron. Greater heat intensity, less ash and a longer burning time required less of it than charcoal extracted from burning wood. The number and distribution of coal-fired iron forges remain unknown. However, Romans transported coal by barges from coal-mining regions to trade for grain and pottery in locations more than 200 miles away by sea and 40 to 75 miles away by land.[2]

Medieval Britain

With the fall of Rome (476 CE), written evidence about coal usage disappeared from the fifth to the eighth centuries. Fast-forward to the Norman Conquest in 1066 CE to 1300 CE. An attack on the woods commenced as a growing population required greater food production. Loggers converted hundreds of thousands of wooded areas, including heaths and marshlands into pastures and farms. The depletion of woods may not have caused a timber famine, but the transportation costs of cutting and carrying cut logs and timber to distant growing cities became prohibitive.

During the reign of Henry III (1216–1272 CE), we have the earliest mention of coal use with deliveries to the town of Billingsgate for heating and cooking. By 1330 CE, enough coal arrived in London, with a population of more than 100,000, to warrant the appointment of coal measurers to calculate coal consumption by the city's residents. In Southwark, Wapping and East Smithfield, coal smoke became a nuisance caused primarily in the kilns of lime burners.

Complaints stated that "an intolerable smell diffuses itself throughout the neighboring places and the air is greatly infected to the annoyance of the magnates, citizens and others dwelling there and to the injury of their bodily health."[3] In the following years, brewers and dyers replaced expensive wood with coal to raise the temperatures in their vats. Metalsmiths also turned to coal in their fabricating activities, all of which added to the complaints about coal smoke on environmental grounds. The energy density of coal when compared to wood, its proximity to towns and cities and its low price gave it a comparative advantage over wood products. By 1340 CE, 1 shilling bought you a chaldron, more than a ton of coal. The monks of Durham priory who owned collieries (coal mines) burned about 200 tons of coal in 1306–1307 CE, and by mid-century, they burned about 300 tons.

The role of coal in the Industrial Revolution

The Industrial Revolution transformed all economies that depended on coal to power its industries. Viewed by most as an inexhaustible energy source, it can easily be cited as an energy revolution. In the preindustrial economy of Britain, including England, Scotland and Wales, coal shipments increased manyfold. Such was not the case for countries on the continent. England's collieries and their productive capacities were unique by any standards. The rapid rise in coal consumption led the statistician, William Stanley Jevons to write about the folly in thinking that coal supplies were inexhaustible. He concluded, "[W]e have to make the momentous choice between brief but true greatness and longer continued mediocracy."[4] According to Jevons, the demand for coal would outstrip the availability of abundant accessible supplies driving up the costs of extraction and adversely affecting economic growth. Fast-forward to the late 1600s, Britain collieries produced 2.5 to 3 million tons of coal, with 90 percent used for homeland consumption. With large peat reserves in urbanized and manufacturing Holland, it burned no more than 65,000 tons of coal annually. The large collieries along the Tyne River produced 100,000 tons a year. Yields in all of France, Germany and what would become modern Belgium produced fewer than 100,000 tons, about 100,000 to 150,000 and possibly a bit more than 300,000 tons, respectively.[5]

After 1700 CE, the British economy passed from dependence on organic sources of power to a mineral source for its energy needs. To a high degree, Britain became dependent on coal during a 130-year period.[6] One can argue that this makes British energy history the proper starting point for understanding the transition from an organic energy regime and one based on fossil coal. By 1700 CE, coal supplied more than half of Britain's fuel needs and possibly that amount as early as 1650 CE.

Britain harvested only a fraction of its woodlands for heating and cooking. Shipbuilding, construction and manufacturing continued as the major consumers of wood products. Growing city populations became major consumers of coal, with fuel-starved Londoners becoming major consumers. Sinking deep shafts to extract the coal and drain substantial amounts of water added greatly to a colliery's production costs. At the same time, transportation costs declined for ships carrying sea coal dug near the sea or navigable rivers and for wagons using an improved network of roads.

The century of the steam engine: powered by wood and coal

Thomas Savery's patent for an atmospheric pump in 1698 revolutionized mine drainage. Called the "Miner's Friend," it allowed operators to dig into deeper and hotter regions. His pump contained a few moving valves in the pump's chamber. There, an injection of steam into the chamber created a partial vacuum by driving out the air. A jet of cold water condensed into steam and filled the partial vacuum with water, forcing it upward into the chamber by normal atmospheric pressure. The injection of steam at pressure was also used to force water upward. Atmospheric pressure can only draw water to a height of 32 feet. So, Savery installed pumps at a greater depth to drain mines.[7]

As noted in Chapter 2, steam-powered engines eventually replaced waterwheels as a new prime mover. First, the ironmonger Thomas Newcomen, in an agreement with Savery, produced a coal-fired steam pump in 1712 CE. It drained water from a coal mine shaft 51 yards deep located at Dudley Castle, Staffordshire. Its boiler held 673 gallons of water. Its 7-foot-tall cylinder measured 21 inches in diameter. Steam from the coal-heated boiler entered the cylinder and lifted the pump's piston.

> Then cold water sprayed into the interior of the cylinder. The voluminous steam condensed to a small amount of liquid, producing a vacuum in the cylinder. Then, came the power stroke, for which all the above was preparatory.

> Atmospheric pressure drove the piston down into the evacuated cylinder. The piston was attacked to a rocker beam, the motion of which could drive a chain of buckets, a bellows, and so on.[8]

In the decades that followed, the cylinders of steam pumps grew to 82 inches to increase their pumping capacity to work at greater depths and at lower costs. In 1769 CE, James Watt, with the assistance of his partner, Matthew Boulton, invented a cheaper steam pump that used about one-third as much coal as the Newcomen engines and one that drained mines from even greater depths. Machinists continued to build and install Newcomen engines in large numbers, despite the work of Watt and Boulton because his engines functioned satisfactorily in draining mines at shallow depths well into the nineteenth century. With mountains of waste coal at the pitheads to power Newcomen's inefficient engines, operators continued to use older technology. Historian Michael Flinn has noted that by 1800 CE, 828 new steam engines of all makes were operational in British coal mines, although the actual number may have been higher.[9] Given the ingenuity of Britain's machinists and engineers, Watt adapted the reciprocal motion of the steam engine to rotary motion in 1781. The transition from steam pumps to steam locomotives became inevitable.

By 1800 CE, the new uses for coal in steam power, iron and gas making accounted for much of its increased use. Fifteen years later, the presence of locomotives in coalfields for hauling mined coal replaced horses. Steam-powered railways reduced transport costs when compared to horse-drawn carriages. Their numbers declined rapidly in the nineteenth century. British mining regions experienced a substantial increase in steam-powered railways decades before the appearance of passenger railways. With improved river navigation and the construction of canals, the coal age flourished.[10]

Industrialization in China

China's coal deposits remained a useful mineral throughout much of its history. Rather than passive recipients of Western technology, Chinese mines used advanced timbering methods to shore up mine shafts and prevent disasters from 475 to 221 BCE. By 1000 CE extensive coal mining supported metallurgy, iron foundries and improved residential heating in northern parts of the country during winter months. The production of pig iron increased six-fold between 750 and 1000 CE. In monetary value, mining was second to agriculture only. The incentive to industrialize during the Qing dynasty (1644–1912 CE) turned its coal into essential power resources to compete in the global market for goods and services. To achieve a renewed status of wealth and power, Chinese bureaucrats, working with observers and investors from Europe and aided by missionary translators, brought the new science of geology to fledgling Chinese mining operations. Large-scale mining enterprises, led by important government officials Li Hongzhang and Zhang Zhidong, made possible industrial expansion. European influence, especially German engineers, brought their acquired skills from mining academies to China's modernizing coal enterprises. By the late nineteenth century, Germany leased mining operations in China's northern Shandong Province.

European empires and multinational corporations entered the market for China's vast coal reserves by getting an imperial foothold in the country. Corporations also engaged in the lucrative business of selling expertise and modern mining equipment. Ownership carried considerable financial risks by investing capital for the modernization of China's mines and iron foundries. The Qing government used state intervention and subsidies to guarantee some level of independence from foreign investment. By the end of the nineteenth century, the government reformed its mining laws in a nationwide effort to deter German, British, French, American and Italian governments and corporations from further exploiting its mineral wealth. Ironically, new mining laws coincided with the weakening of the centuries-long Qing dynasty, its loss to Japan in the Sino-Japanese War in 1905 and its collapse in 1912 caused by provincial disputes over railroad rights.[11] China's Nationalist government that replaced the dynasty was weakened by Japan's invasion of Manchuria in 1931 and its seizing the northern region's coal mines and steel mills. Japan's expansion along China's eastern coasts, World War II and the communist victory in the civil war ending in 1949 led to a decline in coal production.

The age of coal

Locating coal below the surface of the land required numerous "hit and miss" approaches. Despite a few scientific treatises that suggested the composition of the subsurface, miners provided most of the knowledge about soil composition and its various substrata. With the discipline of geology in its infancy, miners provided clues as to the possible location of coal deposits. In searching for outcrops of coal near the surface, simply digging into the topsoil or cutting shallow trenches uncovered seams. Discovering deeper seams required digging an exploratory, costly, labor-intensive shaft.

The invention of boring rods in the early 1600s CE uncovered coal at depths unimagined before its use. As historian John Hatcher has described, the techniques and tools involved a patient commitment on the part of investors, coal merchants and workers:

> Basically it was a percussive/rotative process, during which wrought iron rods were bounced on a beam supported on a timber tripod. At each stroke the handle gripped by the borers was twisted a quarter turn. A chisel of the type appropriate to the substance being cut through was attached to the end of the bottom rod and, as the hole deepened, further rods were attached. Every four to six inches the rods were lifted to enable the chisel to be sharpened and for samples to be taken in a device called a wimble, which collected fragments of rock.[12]

Sinking a shaft to uncover a coal seam represented a more costly alternative, requiring the energy of diggers using picks and shovels without the benefit of equipment other than a windlass installed across the hole to lift the uncovered soil and rock. In the new age of coal-fired steam engines, human and animal energy and timbers remained a significant source of power. The caloric energy of men and horses powered the windlass to lift the shaft's waste. As sinkers (diggers) dug deeper, timbers lined the 6-foot-wide shaft to prevent the sides from collapsing, burying the sinkers who worked either alone or in tandem. As the depth increased, labor and timber costs (employment of carpenters, a ready stock of nails, etc.) increased as well. Sinking through stone and encountering groundwater from underground streams required digging diversionary shafts and/or constructing watertight walls of wood. Usually fir swelled when wet, and either clay or caulk sealed the joints when exposed to water, all of which delayed progress and added to the cost of locating coal seams.

Once located, mine operators used two methods to determine the amount of coal that could be removed. One method, longwall mining, required the use of mine debris and timbers to shore up a mine's ceiling. In a second method, massive coal pillars of undisturbed coal were left in place as hewers (coal miners) hacked away at a seam with picks, hammers, and wedges. Again, historian John Hatcher describes a miner at work:

> Working in confined spaces, the hewer would begin by making cuts in the seam by using his pick. This was a job requiring skill, athleticism, and stamina, for the hewer had to work lying on his side. When this had been done, he would then make a series of vertical cuts along the face of the coal, thereby cutting the coal on three sides. Then the hewer drove his wedges into the coal and, sometimes with the assistance of a crowbar, brought it down in as large pieces as he could.[13]

The human labor expended to get access to mineral-based energy powered much of the early modern world but at what costs to those in the mines? As Hatcher's description suggests, dislodging large pieces of coal, locked in place for millions of years, was back-breaking, lonely but lucrative, by seventeenth-century standards. Hewers worked in dark, damp and physically confining spaces. Since *the Compleat Miner* and *The Compleat Collier* (1709) focused on the works rather than the words of the colliers, whose numbers grew to 1.25 million digging 288 million tons of coal in the industry's peak year of 1913, we possess few testimonies about their work.

Replacing charcoal-fired furnaces and water-powered turbines to power steam engines with coal eliminated the constraints of the former. No longer would manufacturing prosper when located near woodlands or on rivers with uninterrupted streamflow. In Britain, with coal deliveries by sea, along navigable rivers and on improved road and rail networks, large-capacity blast furnaces created economies of scale. They burned at higher temperatures and made an expanded range of products. It began in 1709 CE, when Alexander Darby built the first blast furnace in Coalbrookdale to smelt iron-using coal as a replacement for charcoal. His innovation created a huge new market for coal and one not limited to the seasons of the year. Later in the century, ironworkers used coal to convert pig iron into bar iron as the raw material for ironwares. "Cast iron became a characteristic product of the industrializing and urbanizing economy: from the narrow base of cannon (in war), the use of cast-iron proliferated: pots and pans, grates, stoves and ranges, pipes, beams and structural cast iron, rails and castings for engines."[14]

To accommodate the increased use of coal, replacing wood, in heating and cooking, required newly designed chimneys, flues and grates for the proper ventilation of heavier coal smoke. Narrower hearths and flues, higher coal baskets and taller chimneys improved ambient air quality but blackened the skies of nineteenth-century cities, causing citizens and public health officers to complain about poor air quality. Many long-standing industries, such as lime burning, salt boiling, iron working and textile dyeing, consumed a larger percentage of the 60 percent of mined coal. Salt boiling consumed more coal in the decades after 1700 CE when it required 250,000 tons. By 1830 CE, it required 350,000 tons.

The growth of the iron industry replaced salt as the greater consumer of coal. New industries added to the industrial mix dependent on coal, including soap and sugar boiling and paper manufacturing. The smelting of lead, copper and tin, as well as the making of glass, bricks and tiles, added marginally to the total. Wood, meaning charcoal, remained the dominant fuel source in the metal industries until 1750 CE:

> An output of c. 30,000 tons of charcoal iron in the mid-century probably required 650,000 acres of renewable coppiced woodland: 800 cubic feet of wood (with an energy quotient of 2.5 tons of coal) being required to smelt 1 ton of pig iron."[15]

As these figures suggest, the depletion of woodlands for charcoal production would eventually stymie industrial expansion. Local woodlands would become depleted first and those afar would entail prohibitively expensive transportation costs. "To produce 10,000 tons of charcoal, 40,000 hectares of forest were needed. The maximum range of transport for friable charcoal by cart was 3–5 miles."[16]

The demands for ironware in an increasingly industrial and urban economy could not be met by the energy performance of charcoal. Domestic ironworks produced only 32,000 tons of pig iron, mainly using charcoal, in 1770 CE. Based on coal 30 years later, it was 156,000 tons and by 1830 CE reached 1 million tons. Despite the increasing demands for coal, prices dropped twice as fast as the decline in prices overall. In addition, the opening of new mines and the increasing use of steam power to drain the mines, along with the development of a renewed infrastructure of roads, canals, rivers and railways, bolstered the economy greatly.

After 1750 CE the mining of coal in Britain soared and by 1800 CE by burning 1.25 to 1.5 million tons of coal in steam engines rose to 4.5 to 5 million tons 30 years later. An additional 500,000 tons were used for manufacturing coal gas for consumption. Manufactured gas is the subject of Chapter 5. Despite these increases, more than 80 percent of Britain's coal use produced heat and steam power. Yet, it remained secondary to the power produced using the energy of water, wind and animal power. Only by 1850, did steam power become a major energy source for the country's locomotives, creating a network of industrial and urban centers. Iron production tripled in the decades to follow while coal output doubled and doubled again.[17]

Between 1830 and 1913 CE, coal fueled the expansion of the railway system that made the inland coalfields accessible to growing steam-powered iron ships. As the world's shipping industry converted from sail to steam, coaling stations flourished around the world, many of them flying British banners and providing coal for mostly British ships. They supported the network of international markets of British manufactured goods and its export trade in coal. As Britain's distinguished statistician W. Stanley Jevons, noted in 1865 CE, "[c]oal alone can command in sufficient abundance either the iron or the steam; and coal, therefore, commands this age-the Age of Coal."[18] Steam coal dominated the international market until 1900 when coal exports from the United States and Germany challenged Britain's superior position. The growth of the labor force in the mines reflected the country's position of leadership from 1830 CE until the early twentieth century. Colliery workers numbered an estimated 109,000 at the beginning of this period and expanded to 1,095,000 in 1913. Their real wages adjusted for inflation rose by 70 to 80 percent between 1850 and 1900. For hewers, real wages rose 110 percent.[19]

Hewers spent their working hours bend over into spaces 2 feet high or lying on their sides in a few inches of water. Conditions changed little for more than a century, despite the passage of the Eight Hour Act of 1908. As H.S. Jevons pointed out in 1915, hewers spent eight hours without relief of strenuous physical labor. With pick and shovel, he would struggle through pools of black slush at the end of a shift caused by water percolating from the mine's roof.[20] Assuming that a collier would produce 200 tons of coal working 270 days a year, a labor force of 152,000 would be necessary in 1830. Getting it from the pithead to the consumer may have equaled or exceeded that number. Comparatively, 375,000 to 400,000 worked in the textile industry in 1831 CE.[21] A physician described the life span of a collier:

> About the age of twenty, few colliers are in perfect health, more or less affected with difficulty of breathing, cough, and expectoration. Between twenty and thirty the men declined in strength. In the forties there is a rapid decline in health. After they turn forty-five or fifty they walk home from their work almost as cripples, often leaning on sticks, bearing the visible evidences of overstrained muscles and overtaxed strength.[22]

While others enjoyed the benefits of fossil fuel, mining remained a most dangerous occupation in which the threats and the reality of death and injury were ever present. There were the frequent strikes, layoffs and the debilitating effects of breathing sharp-edged microscopic particles of coal dust that British miners called "black spit" and that we know as black lung disease (pneumoconiosis), tuberculosis, arteriosclerosis, cancer and ulcers. Mine fires, cave-ins, explosions, and electrocution threatening injury and death to miners and the livelihood of dependent families created a culture of anxiety and fear. The failure of the government to protect them caused resentment and hostility to company practices. Heavy and dangerous machines used in cramped, poorly lighted and dusty spaces brought their own dangers.

Mechanization came to the mines slowly, first as compressed air motors used to undercut the coal by a depth of 3 to 4 feet in the 1850s in preparation for dangerous blasting with gunpowder or dynamite. Well established in the larger collieries by the 1880s, small movable electric bar and disc cutters entered the mines to compete with compressed air motors to undercut coal seams. Chain cutters operating as a coal saw possessed the strength of disc cutters and the flexibility of bar cutters in the 1890s. Such inventions reduced dependence on the hewer's muscle power swinging a pick and lifting a shovel full of coal into a wagon or railcar. That mechanization occurred slowly reflected the difficulty of transmitting compressed air and electricity belowground safely. Yet, by 1913, electricity provided the horsepower for 62 percent of the machinery in the mines.[23] Not until 1925 CE did machines dig about 20 percent of the coal. By 1950, 80 percent of coal was machine cut, reducing the workforce. Changed technology did not necessarily improve working conditions, however. As the novelist and social critic George Orwell discovered in 1936 when he went down into a pit: "In Yorkshire, he saw a chain cutter at work. As the machine works it sends forth clouds of coal dust which almost stifle one and make it impossible to see more than a few feet."[24]

Digging coal changed very little during the nineteenth and twentieth centuries, when motive power was either human or horse. Human muscle power hauled the tons of coal to the surface each day. Boys, girls and the wives of hewers stumbled along the narrow and low-ceilinged passageways, carrying on their backs baskets filled with upwards of 100 pounds of coal. Once they reached a mine's eye, its opening to the surface, they climbed up a series of ladders to dump their baskets. As early as 1793 CE, one mine owner called for "a Stop to the *barbarous* and *ultimately expensive method* of converting the colliers' wives and daughters into *beasts of burthen*, and causing them to carry coals to the pit bottom or to the banks on their backs."[25]

In many large mines, girls, wives and boys hauled coal on sledges from the face of the mine to a main underground roadway, where it was transferred to horse-drawn wagons on either a wooden or iron railway. Mine operators used horses aboveground to turn the windlasses that drained mines of groundwater. However, underground, many thousands of horses hauled wagons of coal on wooden and iron railways. Once horses entered the majority of the mines, they remained there for the rest of their lives, with the driest places belowground serving as stables. Replacing horses with steam power in winding after 1790 CE in the larger collieries represented a major technological innovation.[26]

Figure 3.3 Young women mine workers in England, 1842.

The Mines Act (1842) forbade women and children from working underground. The Coal Mines Regulation Act (1860) restricted children's employment aboveground to age 12, 10 if they could provide evidence of formal schooling. In many large mines, girls, wives and boys hauled coal on sleds from the face of the mine to a main underground roadway, where it was transferred to horse-drawn wagons on either a wooden or iron railway. The 73,000 horses and ponies allowed by the Regulations Act provided substitute labor. Mechanical power in the form of stationary steam engines underground and the manufacture of wire rope for winding and hauling replaced horses in the larger collieries.[27]

Peak production of British coal surged from 1870 to 1913 CE from 110 million tons to 290 million tons. By 1900 Britain exported 25 percent of its coal, reaching 73 million tons in 1913. Between the two World Wars, domestic consumption, including heating and cooking, iron manufacturing and cotton textiles, declined slightly with an increase in electrical generation. A precipitous drop in exports, falling from 83.3 million tons in 1924 to 48.7 million tons in 1938, damaged the coal industry irreparably. Declines continued during World War II and dropped sharply in the 1950s, with oil, natural gas and nuclear electrical generation pushing coal production to the side.[28]

At the same time, older British mines faced increasing competition from the United States and Germany, where mechanical cutting replaced manual labor more rapidly. In 1910, machines cut 25 percent of the coal in the United States while a "negligible fraction of British coal" was cut this way.[29] To counter this argument, where mechanization was suitable in larger collieries,

> between 25 and 33 percent of coal-faces located in conditions suitable for machinery were mechanized; this represents an effective rate of mechanization between three and four times higher than that implied by the widely quoted official statistics – and comparable with the rates of diffusion in the United States.[30]

Despite the increased tonnage produced by British miners during this period, the country's share of global output declined as the expansion of the iron and steel industries in Germany and the United States demanded massive increases in the domestic production of coal. The rise in coal production in Russia, Japan and Australia further reduced Britain's predominance in coal production. Under these conditions, maintaining an 85 percent share of the world's trade in coal in 1900 seemed unlikely.[31]

Chinese miners

While few records exist about the work of Chinese miners in the nineteenth century, twentieth-century coal miners faced a fate similar to British and American miners during an earlier century. The nationalization of coal mining by the communist government in 1949 accelerated the migration of the country's vast numbers of agricultural workers into mining where wages exceeded that of farm labor. Wages amounting to US$15 a month for miners improved their social and economic status. The human cost remained incalculable. Miners descended into the stiflingly humid air and darkness of mines, bare-chested. Since explosives dislodged large coal seams, miners carried shovels to move chunks to baskets and conveyers. Thirst, hunger, injuries and death became commonplace.

As state-owned mines expanded and flourished as government money provided the financial support for its industrial expansion, millions of Chinese entered the mines leaving the countryside. Shoveling coal by hand continued despite improvements aboveground with community hospitals, schools and recreational centers available to workers and their families. Descending into mines now took an hour or more as blasting opened coal seams required more monotonous, back-breaking work in the most air polluted of conditions. Workers described the conditions as similar to a bird with its wings clipped. Work never ended, and no matter how hard one worked, one could never fly away.

Wood- and coal-generated steam power

Freedom of location, as noted earlier, would lead eventually to the replacement of wood with coal as the energy source, first in Britain and later in the United States. There, in the decades of the 1850s and 1860s, steam replaced waterpower. As Louis Hunter has noted, "[e]ngines were complex assemblies of fixed and closely articulated moving parts whose fabrication was dependent upon the slowing emerging metalworking crafts: filers, filters, and, above all, machinists, working with a precision and unheard of in traditional crafts, clock making apart."[32]

The first steam engine imported to the American colonies occurred in 1754 CE to drain the Schuyler copper mine in New Jersey. With abundant virgin forests in the United States, unlike Britain and the depleted forests of China, rural environments, close to rivers where waterwheels powered existing machinery, became home to most of the new nation's fledgling manufactories. The transition to steam engines, late by British standards, used wood as fuel. Louis Hunter explained the delayed transition as a function of eighteen and nineteen-century attitudes among mill owners and operators: "From time immemorial the power employed in industry had been identified with falling water and splashing mill wheels. For generations the first step in any industrial venture requiring power was the location of a site with a water privilege."[33]

Britain became urban, industrial and mechanized earlier and experienced the transition to steam power primarily for draining flooded coal mines. In the United States, these interrelated processes began to accelerate in the 1840s. With the rapid growth of many cities, through immigration and natural population growth, and away from falling water, a great expansion of mechanical energy using steam engines for motive power became a requirement for rapid industrialization. From 1840 to 1880 CE, the population in the United States nearly tripled, with the urban share rising from almost 2 million to 14 million, with a percentage change from almost 11 percent to close to 30 percent.[34] The number of industrial workers increased more than five times, and with stable prices, the value of factory goods rose 13 times. Initially, factories built along waterways with the mechanical power of waterwheels replaced human and animal muscular energy. With the urbanization of industrial work, the increasing power of steam boilers fed by wood released factory owners from the limitations of location and the unpredictability of running water.

As one of the U.S. commissioners to the Vienna International Exposition of 1873, Robert H. Thurston, reported that

> the location of mills and manufactories was determined by the availability of mill-privileges, rather than by proximity either to the source of supplies or to the market. All of this is now changed by the marked reduction in the costs of fuel for and maintenance of steam engines.[35]

With the urbanization and industrialization of the nation's economy, an interconnected railway system replaced waterways linking the East with the trans-Appalachian West. Steamboats, along the inland and coastal waters, came first, only to be superseded by railways in the latter decades of the nineteenth century.

As Louis Hunter has noted, the steam engine with the assistance of a furnace that burned wood and then more energy-intensive coal and a boiler that converted water into steam was industry's first practical heat engine. At mid-century in 1850 CE, fuelwood supplied 90 percent of the nation's heat-energy requirements. The rise in steam engines efficiency and power coincided with the growth in trade, urbanization and complex transportation routes that fanned out along waterways and then railways. At the end of the century in 1900 CE, coal supplanted wood almost entirely, supplying three-quarters of the nation's energy needs with wood dropping to less than one-quarter. The transition to a mineral energy regime reflected the change from inland waterways to rail traffic and the replacement of waterpower with steam, first using wood as part of a continuing organic energy regime to mineral energy regime of coal. The transition can be seen in the rapid development of steamboats and their demise.

Steamboats

Applying steam to powerboats followed the practice of draining copper mines in New Jersey. Since the British banned the export of plans to build steam engines using the patent held by Matthew Boulton and James Watt, an American tinkerer, John Fitch (1743–1798 CE), with the financial help of clockmaker Henry Voight, launched a 60-foot paddlewheel boat in 1790 on the Delaware River, serving passengers from Philadelphia to a number of cities and towns in New Jersey. Powered by burning wood to turn water into steam, its paddlewheel turned at a rate of seventy-six strokes a minute. The commercial venture failed as expenses exceeded revenues. This failed venture left the field to Robert Fulton (1765–1815 CE) and his influential political and business partner, Robert Livingstone (1746–1813 CE).

After Fitch's death by suicide, the New York State legislature transferred his rights to operate steamboats on the Hudson River to Livingstone. The inventor, Fulton, and the entrepreneur, Livingstone, met in London and agreed to build a steamship. It was Fulton, however, whose technical skills in munitions aided the British navy. For this, he received payment and the rights to export a Boulton-Watt engine. At the East River ship works, Fulton launched the *Clermont*, named for Livingstone's sprawling Hudson River estate, on August 17, 1807. Fueled by coal from Pennsylvania fields, rather than wood from upstate New York, this 146-foot-long, 12-foot-wide paddleboard ship with a 15-foot smokestack set sail for the two-day journey up the Hudson River to Albany, New York. With the ship traveling at about 5 miles per hour, the return trip took about 30 hours.

By investing all profits in additional and improved steamboats, by 1812, ferry service with 21 boats traveled along the Hudson, Delaware, Potomac and James Rivers, as well as the Chesapeake Bay. Undeterred, both men planned to build steamboats that could travel the inland waterways from industrial Pittsburgh to commercial New Orleans. They reached out to Nicholas Roosevelt (1767–1854), an engineer who invented the vertical steamboat paddlewheel to build the 310-ton *New Orleans*. Traveling the 2,000-mile route, the steamboat left Pittsburgh in October 1811 and reached its destination in January 1812, traveling at a speed of 8 miles per hour and experiencing an earthquake and an attack from First Nation warriors. There, it remained, ferrying passengers and products from the Gulf of Mexico to Natchez, Mississippi, before running aground near Baton Rouge,

Louisiana, in 1814. In the next quarter century, steamboat travel became global. In 1838, the Steam Ship (SS) *Great Western* crossed the Atlantic Ocean, and on October 6, 1848, the SS *San Francisco* steamed from New York Harbor for San Francisco, around Cape Horn, arriving on February 28, 1849. Humans no longer depended on wind and sail to cross Earth's seas and oceans.[36]

Although the names of Fitch, Fulton and Livingstone dominate discussions of the importance of steam power to move people and goods in trans-Appalachian West, Louis Hunter reminds us that "[t]he steamboat, like practically every mechanical complex of importance, was the product of many men working with a common heritage of technical knowledge and equipment and impelled by a common awareness of need."[37] It should be noted, however, that Fulton, among others, believed steamboats provided the cheap and quick way to carry people and product along the Mississippi, Missouri and other western rivers. With woefully poor roads throughout the new nation, cheap and efficient river transportation provided a necessary ingredient for economic development and westward expansion. The era of building larger and more powerful steamboats capable of traveling downstream to the Gulf of Mexico port cities and returning upstream with their hulls filled with cargo commenced with enthusiasm and much skill to meet the demands of the country's growing population and its vibrant economy.

Within a few decades, traveling 100 miles a day became commonplace. Delays caused by fuel shortages at wood yards along the way or for disruptions in felling, cutting, stacking and delivering wood to way stations along the river courses lengthened travel times. As Julia H. Latrobe noted on her trip from Pittsburgh to New Orleans in 1820,

> We had to stop for wood almost every day; this is a great thing for the country people and new settlers, for as they clear away the land for their fields, they cut the wood and pile it along the bank, where it is in constant demand from the steam-boats.[38]

At mid-century, fuel costs for cordwood became excessive. In the 1850s, loggers cleared 50 million acres to produce 75 million cords of wood. Although much of it was used for heating and cooking, increased cordwood consumption by steamboats began making coal costs attractive. A steamboat run from Louisville to New Orleans burned on average 529 cords of wood at the cost of US$1.25 to US$6.00 a cord. Coal cost half as much. Ten to 12 bushels of anthracite coal equaled a cord of wood, measuring 4 feet high, 4 feet wide and 8 feet long, or 128 cubic feet. Pushing hard to remain on schedule, a large paddlewheel steamboat could burn 30 cords a day, making a number stops along the way to refuel with wood supplied in many places by wood hawks.[39]

Transitioning to more affordable coal, which also produced much more energy by volume than wood did, did not save steamboats from competition from railways that began to link the East with the trans-Appalachian West. The golden age for steamboats ended in the 1850s as railways penetrated markets, gaining direct access to raw materials and finished goods without the limitations imposed by waterways. "The agencies of river transport operated within the fixed and rigid framework provided by nature and were unable to extend materially the territorial range of their service beyond service early reached."[40]

Fuel costs favored substituting coal for cordwood. Although the early railways experienced their share of collisions, fires and breakdowns that disrupted service and caused expensive delays, fires and the breakage of steamboat gearing, heavy losses caused by steamboat disasters, including boiler explosions, collisions, fires and groundings, tilted the competition in favor of rail traffic.

Although the role of steamboats as a vital part of the new nation's growing transportation system was an abbreviated one, its legacy could be found in the steam engines that mechanists built that composed as much as three-fifths of the nation's steam power. They included the machine shops, where workers tinkered to increase the power of these large and somewhat cumbersome engines, and the foundries and boiler works that increased the nation's industrial capacity. These metalworks facilitated the transition from wood as a fuel source and as a building material to iron for rails, locomotives and structural components.

With the passing of its golden age and the loss of its primacy in shipping and passenger travel on the nation's rivers, steamboat service declined but did not disappear in the post–Civil War years, 1865 and beyond. Paradoxically, some of the industry's largest and most luxurious steamboats became operational during these years. Steamboats of the 1,000-ton class capable of carrying cargo twice that amount plied the waters of the Mississippi–Missouri watershed in the 1870s. To attract wealthy passengers, the largest of these vessels decorated cabins with stained-glass skylights, rich carpets and elegant furnishings. Despite these efforts, steamboats greater than 200 tons declined by 50 percent from 1870 to 1890, while owners and operators of lower-tonnage vessels used them for short-distance travel. Contracting further with the growth of the railroads, the era of the steamboat ended in the late nineteenth and early twentieth centuries with the arrival of internal combustion engines in trucking and diesel-powered tugboats pushing coal- and iron ore–laden barges to industrial mills along the rivers.

The global transition to coal

While the Lowell, Massachusetts's textile mills powered its factories with the energy of flowing water; they heated their facilities with anthracite coal. By 1833 CE, they imported more than 7000 tons for this purpose.[41] The overlapping existence of organic and mineral energy regimes functioning simultaneously was commonplace. As noted earlier, picks, shovels, wedges, hammers and wagons pulled by boys, girls and women and horses would dislodge coal from its Carboniferous origins 300 million years ago. Human and animal muscles, representing an organic energy regime, built and developed by metabolizing food from plants and animals would provide the energy to work belowground for centuries.

Historian Christopher F. Jones suggests that definitions vary when describing the phrase "energy transition." He noted that, "in some cases, a transition means abandoning a former source of energy, while in other cases a new energy source is added to the mix without eliminating earlier practices"[42] aptly describes the process of transitioning. As the United States became more urban and industrial, the nation's abundant forests used traditionally for construction materials, furniture, heating and cooking witnessed an accelerated urban demand for timber and the new demand from factories requiring cordwood to provide heat and energy to power their machines.

At the beginning of the nineteenth century, cordwood supplied the major industrial and transportation needs of the country as the steamboat era suggested. Even though it required literally thousands upon thousands of hours to cut, split, stack and transport, to say nothing of the months required for seasoning it before burning, abundant and available wood would remain the nation's primary organic energy resource. Once transportation costs began to rise, caused by disappearing proximity and the accelerating pace of industrialization and urbanization, its use dwindled. Cordwood (4 ft × 4 ft × 8 ft) consumption peaked in 1870 CE at 1,406,985 cords and fell rapidly thereafter as the nation shifted to coal.[43]

The transition from wood to coal followed a pattern seen earlier in Britain. It happened slowly, with some regions of the country making the transition earlier than others. By 1800 CE, Pittsburgh's early industries belched clouds of black smoke from burning available soft, bituminous coal. The city "could be identified from a distance by the heavy pall of coal smoke that lay over the city."[44] With coal consumption from all uses approaching 400 tons daily in Pittsburgh in the 1830s, mines near the city produced as much as 200,000 tons annually, with its ironworks using annually two-thirds of 75,000 tons of coal used for industry. "Coal has been the life of the steam engine, and the steam engine has been the great power which has called into existence our manufacturers."[45]

With a typical household of six people in 1820 burning a bit more than a cord of wood per person each year, Philadelphia's population of 137,097 in 1820 consumed 175,000 cords of wood. When combined with the totals for New York City, the number reached 600,000 cords, with Philadelphia industries burning another 100,000 cords.[46] At this rate of cutting and burning wood, its cost increased as a function of the distance required to cut and transport wood to market. Although the nation in the early decades of the nineteenth century possessed millions of trees, loggers had already thinned those close to urban markets. With costs rising, coal became a competitive choice in industries dependent on the power of steam and households and factories needing a source of heat for warmth. A cord of wood could not compete with the heat value of a bushel of coal.

The rewards and benefits of burning anthracite coal were not shared equally. Synergistic feedback loops sustain mineral-based energy regimes. Cities attract workers to their factories and mills; continuous flows of mineral energy attract investment that attracts more workers, producing more products. Increased mineral flows never exceed the law of diminishing returns in which land used for timber interferes with land needed to feed a growing population. As Jones pointed out, networks to increase mineral energy flows in the form of canals, roads and bridges were necessary to expand the delivery of coal.

From 1850 CE coal replaced wood rapidly, with coal production doubling every 10 years. Its virtues were proclaimed in the most expansive terms. In not-so-veiled racist terms,

> coal was not only the basis of the industrial and maritime power of England but, more comprehensively, the cause of the superior character of the coal-using inhabitants of the northern hemisphere to that of the dwellers in the southern and tropical latitudes.[47]

Before the transition from wood and charcoal to coal, wood fuel provided 90 percent of the heat energy requirements of the United States as late as 1850. Burning wood to make charcoal for manufacturing purposes, felling, cutting, splitting and stacking cordwood was a self-sufficient family activity. Gaining access to coal beds, however, required major capital investments, including the development of the transportation networks described earlier. Miners found few deposits close to the surface. Most coal beds required muscle and animal power for digging, hoisting, draining and ventilating the mines. The invention of steam engines relieved some of the pressure from humans and animals, but in an industry that mechanized slowly, hewing coal required a labor-intensive, physically exhausting, dangerous and depleting endeavor.

As noted earlier, during this 50-year period, urbanization, industrialization, railways, population growth stimulated by massive immigration from Southern and Eastern Europe to provide workers for the nation's factories and mills required a high-energy, easily transportable fuel source. Coal burned at a much higher intensity than wood, weighted less and cost less to move from pithead to commercial and residential establishments. Unlike wood, where its depletion was self-evident, coal belowground seemed inexhaustible.

After 1850, the increasing production of coal mines in Ohio, Indiana, Illinois and Missouri, along with the growing network of railroads, opened the region to manufacturing and a shift from rural sites requiring waterpower to towns and cities that linked raw materials, factories for production and markets for consumers. As Louis Hunter has pointed out, as the nation expanded westward after the Civil War (1861–1865), the use of anthracite and bituminous coal grew by millions of tons annually. Leading ports at Buffalo, Cleveland, Detroit, Milwaukee and Chicago, with growing populations and expanding manufacturing capacities, contributed greatly to the use of coal's thermal energy. Moving farther West, the coalfields of Colorado, Utah, Wyoming and Montana fed the energy needs of the regions metal-mining industries. The extension of the railroads and their fuel source, coal, contributed to the productivity of the region's copper and silver mines.

By 1880 CE, wood, thatch, straw and other biomass gave way to coal as the world's main commercial energy source. By 1885, locomotives consumed two-thirds of the nation's output of bituminous coal.[48] At the same time, most of the world's coal supply existed in a few countries where it was consumed for domestic consumption and for industrial production. The combination of coal and steam power revolutionized the production of capital goods, iron and steel. Mine owners produced so much coal that the supply increased as much as 50 percent each decade and its use in industry required access to the raw material resources of nonindustrial countries capable, for example, of producing large quantities of sugarcane and bales of raw cotton.

Forged in blast furnaces fired by coal using steam-driven bellows, steel rails and locomotives connected concentrated urban populations and brought finished material goods to rural areas. Iron- and steel-framed ships burning coal in their boilers powered oceangoing vessels, replacing the clipper ships powered by the wind captured in its sails. Coaling stations located strategically around the globe serviced this new generation of ships engaged in transnational trade bringing raw materials to Britain, its European neighbors and the United States. Imported labor in slaves and indentured servants supplemented local workers. Formerly, self-sufficient nonindustrial countries became dependent colonies supplying fossil fuel–intensive countries with its commodities, precious metals, cotton, sugar, rubber, tin and many others. In the transition to dependency, self-sufficient farms became single-crop-cash economies.

Households used fuelwood primarily as kindling to ignite anthracite in a stove or firebox as late as 1890. By then, coal had replaced wood in cities, factories and in transportation. Much like the earlier fuel transition in Britain, coal had become the ubiquitous fuel, available at a cost cheaper than wood, more easily transportable, and with greater heat energy by volume. By 1900, self-sufficiency in meeting one's heating and cooking needs ended in the United States. By that time anthracite and bituminous coal, the more plentiful and softer coal, provided about three-quarters of such needs. Wood supplied about one-fifth. With these fuel sources providing about 90 percent of the need for heating and cooking, peat and biomass filled the remaining gap. Pittsburgh, in the nineteenth and early twentieth centuries, became the country's largest producer of bituminous coal.

The Pittsburgh coal seam, covered a 14,200-square-mile area, covered parts of 12 Pennsylvania counties and bordered New York State. The development of the region's rail system delivered coal to markets in Pennsylvania and beyond. The best year for delivery by rail was 1918, when it carried 128,518,000 tons of coal to industry and to homes. By comparison, barges along the region's developed river system carried 8,985,000 tons to market. Beginning with the 1920s, barge traffic, cheaper than rail at about 18 percent of the railroad rate, began to catch up. With sharp drops in coal consumption caused primarily by competition from oil and natural gas, barge traffic excelled when coal could be transported from pit to industrial mills and utilities, burning coal to produce electricity.[49]

In the American West today, the energy to provide light, heat and electricity comes now from coal-burning power plants supplied by enormous strip mines in the Powder River basin of Wyoming and the mostly mechanized deep-shaft mines of western Colorado and eastern Utah. The history of the transition from wood to coal begins not in the American West but in the anthracite coal mines of eastern Pennsylvania. Historian Christopher Jones has pointed out that entrepreneurs first needed to build a transportation network by dredging rivers and cutting canals to deliver coal mined from its primary location in the Lehigh Valley of Pennsylvania to consumers at a reasonable cost and then to convince consumers to make the transition from wood to anthracite coal for cooking and heating. Before the completion of these significant infrastructure developments, the cost of transporting finished goods across the Atlantic at US$9.00 a ton came close to the transport costs, before infrastructure investments, of transporting coal from pithead to Philadelphia and New York City consumers. Changing consumer behaviors and penetrating the vast consumer market for firewood would require a promotional campaign on the part of anthracite coal entrepreneurs.

The solidarity of miners

As historian Timothy Mitchell has argued, the rise of coal created opportunities for solidarity among workers:

> Great volumes of energy now flowed along narrow, purpose-built channels. Specialized bodies of workers were concentrated at the end-points and main junctions of these conduits, operating the cutting equipment, lifting machinery, switches, locomotives and other devices that allowed stores of energy to move along them. Their position and concentration gave them opportunities, at certain moments, to forge a new kind of political power.[50]

As others enjoyed the benefits of fossil fuel, mining remained a most dangerous occupation where the threats and the reality of death and injury were ever present. In Britain, 94 miners' strikes took place between 1830 and 1888. Wage rates in 87 of these strikes were the primary issue, with 58 strikes called because owners wanted miners to accept a reduction in wages. Owners won only 22 times. In the period between the two World Wars, miners bore the brunt of a depressed industry. They experienced crushing defeats in the 1921 strike and the General Strike of 1926, and many miners' households saw wages fall below subsistent levels.[51]

From the 1880s onward, strikes by workers over working conditions and wages challenged the powers of mine owners. From 1881 to 1905 in the United States, coal miners struck at a rate about three times the average as workers in other industries and twice as often as those in the tobacco industry. It took the intervention of President Theodore Roosevelt to settle the Anthracite Coal Strike of 1902. Comparing coal-mining strikes with those of all other industries, the following statistics, the first number representing strike rates by miners and the second representing strikes by all other industries are suggestive: from 1881 to 1886, 134 and 72; from 1887 to 1899, 241 and 73.3; from 1894 to 1900, 215 and 66.4; and from 1901 to 1905, 208 and 86.9. In addition, coal miners' strikes lasted much longer.[52]

As noted earlier, before mining became mechanized, hewers worked in pairs along coal seams, leaving pillars and walls of coal and rock to support a roof and to separate them from pairs working in adjacent chambers. They worked independently making decisions about where and how much to cut to prevent cave-ins.

"The miner's freedom from supervision is at the opposite extreme from the carefully ordered and regimented work of the modern machine-feeder."[53] According to Mitchell, miners' strikes resonated with other workers because burning mined coal released a flow of energy that connected work belowground to households, factories, offices and transportation networks that depended on steam, and later electricity, to power modern society.

Although this chapter has highlighted energy regimes in China, Britain and the United States, strikes were commonplace in other countries as well. The Zonguldak coalfields on the Black Sea in Turkey were wracked with repeated strikes. Coal heavers at the world largest coaling station at Port Said in Egypt struck in April 1882. In 1890, miners struck in England and Wales, "by far the biggest strike in the history of organized labor" with 260,000 men, women and children demanding wage rates and fair labor standards. Unlike other industries, slowing down energy flows through interconnected industries, starting with coal miners and spread to railway men, dockworkers and transport workers, all depended on those extracting buried "sunshine" from the earth. Uniting in a common cause, they stopped work or slowed it down for increased hourly pay and an eight-hour workday.[54]

The Great Colorado Coal Field Wars that lasted 10 days following the Ludlow Massacre on April 20–21, 1914 represented the culmination of conflict between labor and capital in the western Colorado coalfields that included strikes beginning in 1884–1885, 1894, 1903–1904 and 1913–1914. Miners from different races and more than 30 nationalities arrived in Colorado having traveled by steamship and on railcars to work in dangerous and dynamic settings to satisfy the American West's thirst for energy. Their pugnacious qualities and opposition to unfair labor practices and low wages brought them into conflict with mine owners.

In the case of the Ludlow mines, the United Mine Workers (UMW) was pitted against the Colorado Fuel and Iron Company (CFIC), owned by John D. Rockefeller. Militia brought to Ludlow by CFIC to protect strikebreakers faced with striking miners armed with rifles. The battle that ensued sparked an extended war across Colorado's southern coalfields. Federal troops sent by President Woodrow Wilson (1915–1921) brought the war to a peaceful conclusion, as strikers refused to do battle with federal authority. Having stopped work seven months before, the strikers and the UMW were spent, financially, physically and emotionally. Although federal troops had not intervened on the side of the CFIC, as the Colorado Militia had, it allowed the company to advertise and recruit new workers to the mines. Once the UMW and the CFIC reached a settlement in December 1914, as many as 70 percent of the mine's workforce was composed of new workers. The end of bloodshed and the truce that ensued established the Rockefeller Plan, a company union that allowed miners to negotiate wages and conditions in the mines. Also, it prevented them from joining independent unions, such as UMW.[55]

The UMW, with legislative support, managed to outlaw company-controlled unions. In the Colorado coalfields, strikes in 1919, 1921–1922 and 1927 under UMW's control sought to extend the limited negotiating rights gained after

Ludlow by demanding the eight-hour workday, workers' compensation for work-related injuries and fatalities and more. At the same time, the production of coal worldwide peaked in the years following the Great War (1914–1918). The introduction of gasoline-powered automobiles, trucks and farm machinery led to steep declines in colliery employment:

> Petroleum and natural gas became the main drivers of a Western economy more dependent on fossil fuel-burning technologies that initiated wide-ranging transformations in production processes, consumer aspirations, physical structure and the fabric of daily life in cities and suburbs, and much else.[56]

The shooting wars between miners, strikebreakers, state militias and intervening federal troops occurred across the country from the Appalachian coal regions of mountainous West Virginia and Kentucky to the Ludlow mines of Colorado. Open strife became commonplace and added to the list of injured and deaths. To the hostility that exploded into violent confrontations was added the denial for years by mine operators and the government that mine dust destroyed respiratory function. Even though miners in other countries had received compensation for black lung decades before, it took legal action and pressure from the United Mine Workers of America to get the U. S. Congress to enact the Federal Coal Mine Health and Safety Act in 1969. Within seven years after its passage, 225,000 miners received financial and medical benefits as well as the 140,000 widows whose husbands had died of black lung.

The decline of the coal industry in Europe and the United States

While peak coal was reached in 1910, a decline in consumption had already begun. Bituminous coal consumption rose 1 percent while petroleum rose 662 percent and natural gas 2,589 percent.[57] Until recent years, coal fired about 40 percent of the country's power plants generating electricity for industry and domestic consumers. As power plants transition to cheaper natural gas, the U.S. Department of Energy predicts that the percentage will drop to 34 percent by 2040. In 2013, the United States produced 985 million tons of coal, the first time in 20 years that the figure had fallen below a billion tons. The electric power industry consumed 93 percent, or 924 million tons, with the figure expected to drop in the coming decades. The nineteenth-century boom in coal that watched its market share stabilize and then begin to fall in the twentieth century may have met its match in the twenty-first century, with a cheaper fuel source in natural gas and an increased regulatory environment imposing further restrictions on emissions of greenhouse gases.

At its height in 1913, about 20 percent of the British adult male workforce or one in five males was a miner. They gained and maintained center stage in the country's economic fortunes, responsible for fueling its locomotives and steamships that dominated domestic and international transportation. Coal smelted the iron ore for its iron and steel manufacturing; machinery and pumps depended on it to raise steam power; it generated heat and illumination that limited the effects of seasonal temperature variations and changes in the natural cycle of day and night. For the world in 1913, coal accounted for 75 percent of all energy. In coal-rich Britain, it reached 94.5 percent, with gas and electricity accounted for almost another 5 percent (both occurring by burning coal). Globally, firewood accounted for 17.6 percent of the world's energy.[58]

At the outbreak of the Great War (1914–1918), a robust industry employed 1.1 million colliers producing 269.5 million tons in 1909–1913. By the end of World War II (1939–1945), employment had fallen to 697,000, and production declined to 188 million tons, a 27 percent decline. Coal supply exceeded demand. Much of the coal continued to be extracted by muscle power, but now coal-cutting machines and pneumatic picks extracted about 8 percent of the coal, replacing multitudes of miners. The closing of smaller mines and mechanization that happened slowly led the way as responses to the overall decline in demand. For Britain, machines came slowly to the mines, 8 percent in 1913 at the height of demand, 19 percent in 1924 and 42 percent in 1933.

By comparison, in the United States, machines extracted almost 50 percent of the coal in the 1920s. Alternative fuels, diesel fuel and gasoline, for the reasons cited previously, made deep inroads into Britain's coal regime. By 1913, coal production in the United States exceeded British production by almost 80 percent. However, during this peak year of production, Britain superabundant coal reserves supplied 23 percent of the world's total supply and about 97 million tons used by foreign shipping for international trade.[59]

A postwar boom in exports was followed quickly by increased competition from the United States, a French ban on British coal on the grounds that it exploited consumers and a depression that hit British heavy industry following the hyperproduction output during the war years. Wage gains for miners and profits for the industry evaporated quickly. For miners, the depression proved to be a disaster. Average earnings in 1921 dropped by 43 percent while the cost of living had fallen by only 24 percent.[60]

A postwar slump in the United States led to a sixteen-week coal strike. At the same time, the French occupied the Ruhr coal region of Germany in January 1923. Defeated in the war and dealing with the requirement to pay war reparations for its role in igniting the Great War weakened Germany's Weimar Republic (1919–1933). These events proved to be fleeting, at best, for Britain's miners. Once international competition reemerged, exposing the country's overcapacity, more mines closed, resulting in further job losses and strikes. Economic stagnation became the norm during the remainder of the decade.

Technological innovations in transportation and industry exacerbated the economic malaise that hit the mining industry with ferocity. Between 1914 and 1924, coal-burning ships registered by the insurer Lloyd's of London fell from 89 to 66 percent of the total number of ships, while diesel- and gasoline-powered vessels rose from 3 to 31 percent. In the decades that followed, the gap would grow greatly, with oil powering most of the world's domestic and oceangoing ships. Efficiencies in coal-burning machinery also led to declines in consumption.[61]

The Royal Dutch company built the first oceanic oil tanker, the *Vulcanus*, launched in December 1910, whose steam engine was powered by diesel fuel. It signaled the demise of the coal as the fuel of choice as it had previously replaced wood. The use of diesel fuel also eliminated a robust and organized labor force of colliers, railcar workers, haulers, stokers and others involved in the hewing and transport of coal. Several coal mines in Europe and the United States closed during the Great Depression (1929–1938), with many more closing in the 1950s. The mechanization of the cutting and loading coal in the 1920s further reduced the workforce and changed the former relationship among hewers working independently in the mines. With the transition to petroleum, a fluid that was lighter than coal, its production continued to decline. Since the 1920s, producers exported as much as 60 to 80 percent of the world's oil production.[62]

As Timothy Mitchell has pointed out, the transition to oil weakened worker solidarity. A few workers monitored the pipelines and pumping stations that transported oil across the country. No need existed to load and unload liquid carbon, as was the case with coal.

> In fact, oil pipelines were invented as a means of reducing the ability of humans to interrupt the flow of energy. They were introduced in Pennsylvania in the 1860s to circumvent the wage demands of the teamsters who transported barrels of oil to the rail depot in horse-drawn wagons.[63]

Releasing the stored energy of coal required combustion and armies of haulers and stokers in steam plants and on steamships. Lighter and burning with little residue, petroleum emancipated, in the words of Lewis Mumford, "a race of galley slaves, the stokers."[64]

Despite minor improvements in profits and wages in the latter years of the 1920s, the Great Depression resulted in a precipitous decline in consumption across all sectors of the economy, including per capita use of coal for cooking and heating. By 1932, per capita use had fallen to 67 centum weight (cwt; one cwt equals 112 pounds) per year, the lowest in 32 years.[65] World War II (1939–1945) coal output continued its downward trend aggravated by a manpower shortage and one that was not alleviated significantly by the conscription of young men into the mines.

Britain's nineteenth-century prosperity, with coal providing the energy to catapult it into a preeminent status as a world power, disappeared as demand for its products faced stiff international competition and the transition to oil and the efficiencies gained by industries that continued to depend on coal. The Parliamentary decision to nationalize the coal industry and substitute private ownership for public ownership, a fundamental break with the past, did not significantly change the structural arrangements of mining. On January 1, 1947, Vesting Day, nationalization of the mines became a reality. A public investment to modernize mining was promised and notices through the coalfields announced, "This colliery is now owned and managed by the National Coal Board on behalf of the people."[66]

From 1946 until 1994, coal was a nationalized industry in Britain. Nationalization, however, did not stop the decline in demand for coal. Its survival depended on providing service for a smaller market in electric generation until the 1980s when other power sources eclipsed coal energy and public ownership of the electric industry ended with privatization. The oil crisis of 1973 initiated by the Organization of Petroleum Exporting Countries (OPEC) to curtail production created a temporary demand for coal.

Downward pressure on coal usage continued, despite the investment of 600 million pounds to modernize the industry. Concerns about the environmental effects of burning coal and may have accelerated the downward spiral. The Great London Smog (1952) caused more than 4,000 deaths and led to an energized antipollution control movement in the country. The Clean Air Act (1956) limited residential coal use, resulting in a switch to natural gas. The conversion quickened with the discovery of natural gas in the North Sea in December 1965.

Further declines in coal use occurred as petroleum use tripled between 1960 and 1973 and the first nuclear power plants began generating electricity in 1959. The Great Miners' Strike (1984–1985) failed to stabilize the industry. The end of public ownership in 1994 and the accelerated introduction of natural gas called the "dash for gas" displaced the equivalent of 50 million tons of coal. Between 1990 and 1994, employment in mining dropped from 49,000 to 10,000.[67] Most of country's remaining 170 underground mines closed in the 1990s. In 2015, two of the three deep-pit coal mines closed, Nottinghamshire's Thoresby Colliery and Yorkshire's Kellingley Colliery.[68] The coal mining industry's long and contentious history had come to an end. The downward trend in employment and demand was slow, painful to the many but unmistakable.

China's coal production

In China, coal consumption rose each decade from the 1950s into the twenty-first century from 335 metric tons to 3.2 billion metric tons in 2010. By burning coal, China produced over a billion megawatt-hours of electricity with plans to double its capacity in the coming decades. Sixty thousand mines of all sizes fed the industrial expansion. The expansion led to increases in wages and valued amenities, including the opening of preschools for the children of miners. Job growth in coal mining reached a plateau in 2000 as production exceeded demand. Becoming a net importer of coal from Australia in 2009 did little to create congruity in the supply and demand matrix. Closing one-half of the country's mines caused more than 1 million job losses. In 2016, mine workers protested delays and wage cuts, flying banners that proclaimed, "[W]e must live, we must eat." Months passed before miners received wages.

To solve the mismatch between the millions of miners and declines in productivity caused by overcapacity, Chinese mine operators began to mechanize the mines by purchasing heavy mining equipment from Germany and the United States. Now the coal mines require skilled workers rather than continuing the traditional labor-intensive model. Coal miners in China, as well as coal-producing countries everywhere, deal with the impact of mechanization and the permanent loss of jobs. *Decarbonization* has become the watchword in China as it attempts to reduce coal-produced air pollution that has strangled life in the cities and caused outbreaks of respiratory diseases and premature deaths. As described in chapters devoted to renewable energy sources, solar plants and wind farms have begun the long process of reducing China's carbon footprint.[69]

Conclusion

The transition away from coal will be long and arduous. Yet, the current turn suggests reasons to become optimistic about its reversal. Globally, coal consumption dropped by 1.7 percent while the United States witnessed a 10 percent drop to figures not seen since the 1970s. China continues to burn about one-half of the world's coal production, yet consumption fell 1.6 percent in 2016. This figure reverses China's annual growth in consumption of 3.7 percent in the previous decade. In Britain, it dropped by 52.5 percent as renewable energy generation expanded significantly. With a drop in global growth, carbon emissions stabilized.[70]

While current coal production generates as much as 37 percent of the world's electricity, further declines will be a boon for public health. Global declines will result in statistics like those United States. Since coal-plant pollutants are linked to cancer and cardiovascular, respiratory and neurological diseases, curtailing and/or eliminating fine-particulate pollution will see a decline in these diseases, leading to a healthier global population and a decline globally in death rates caused by burning coal. In India, coal emissions cause between 80,000 and 115,000 deaths; in the United States, 13,000; and in Europe, 23,000. As coal production declines a rise in human health and quality of life becomes possible. "Given the scale and scope of the energy transition now underway, the choices utilities make to replace coal will have a major impact on public health, the environment, and economic justice."[71]

Notes

1 Lawrence P. Lessing, "Coal," *Scientific American* (1955): 59–60.
2 John Hatcher, *The History of the British Coal Industry: Before 1700: Toward the Age of Coal*, vol. 1 (Oxford: Clarendon Press, 1993), 16–17.
3 Ibid., 24–25.
4 William Stanley Jovens, *The Coal Question, an Inquiry concerning the Progress of the Nation, and the Probably Exhaustion of Our Coal Mines*, 3rd ed., revised (London: Macmillan and Co., 1906), 460 quoted in Victor Seow, *The Coal Question* in the Age of Carbon at https://sites.fas.harvard.edu/~histecon/energyhistory/seow.html.
5 John Hatcher, "The Emergence of a Mineral- Based Energy Economy in England, c. 1550–1850," in *Economia e Energia, SECC. XIII-XVIII, Instituto Internazionale Di Storia Economica*, ed. Simonetta Cavaciocchi (Le Monnier, 2002), 483–484.

6 Michael W. Flinn with the assistance of David Stoker, *The History of the British Coal Industry, Volume 2, 1700–1830: The Industrial Revolution* (Oxford: Oxford University Press, 1984), 455.
7 Peter Mathias, "Economic Expansion, Energy Resources and Technical Change in the Eighteenth Century: A New Dynamic in Britain," in *Economia e Energia*, 114–115.
8 Alfred W. Crosby, *Children of the Sun: A History of Humanity's Unappeasable Appetite for Energy* (New York: W.W. Norton & Co., 2006), 73.
9 Flinn, *The History of the British Coal Industry, 1700–1830*, 127.
10 Ibid., 188–189.
11 Shellen Xiao Wu, *Empires of Coal: Fueling China's Entry into the Modern World Order* (Stanford: Stanford University Press, 2015).
12 Hatcher, *The History of the British Coal Industry: Before 1700*, 196–197.
13 Ibid., 209.
14 Peter Mathias, "Economic Expansion, Energy Resources and Technical Change in the Eighteenth Century," 30.
15 Ibid., 29.
16 Ibid.
17 Hatcher, "The Emergence of a Mineral- Based Energy Economy in England, c. 1550–1850," 500–501.
18 William Stanley Jevons, *The Coal Question* (1865), vii–viii.
19 E.H. Hunt, *British Labour History* (London: Oxford University Press, 1981), 73.
20 Ibid., 367.
21 Flinn, *The History of the British Coal Industry, 1700–1830*, 365.
22 Michael Pollard, *The Hardest Work Under Heaven: The Life and Death of the British Coal Miner* (London: Hutchinson & Co., 1984), 51.
23 Ibid., 48–49.
24 .Flinn, *The History of the British Coal Industry, 1700–1830*, 365.
25 Ibid., 335.
26 Ibid., 97.
27 Roy Church, *The History of the British Coal Industry, Volume 3: 1830–1913: Victorian Pre-eminence* (Oxford: Clarendon Press, 1986), 241.
28 Walter Minchinton, "The Rise and Fall of the British Coal Industry: A Review Article," *VSWG: Vierteljahrschrift Für Sozial- Und Wirtschaftsgeschichte* (1990): 219–220 at www.jstor.org/stable/20735640.
29 Peter Mathias, *The First Industrial Nation: An Economic History of Britain 1700–1914* (London: Methuen & Co., 1983), 378.
30 Church, *The History of the British Coal Industry, Volume 3: 1830–1913*, 770–771.
31 Ibid., 772.
32 Louis C. Hunter, *A History of Industrial Power in the United States* Volume Two: *Steam Power* (Charlottesville: University of Virginia Press, 1985), XXI.
33 Ibid., 90.
34 U.S. Bureau of the Census, *Historical Statistics of the United States, Colonial Times to 170*, pt. 1 (Washington, D.C., 1975), series A 43–56, 11.
35 Hunter, *A History of Industrial Power in the United States Volume Two*, 116. fn.71.
36 Michael B. McElroy, *Energy: Perspectives, Problems, & Prospects* (New York: Oxford University Press, 2010), 234–237.
37 Louis C. Hunter, *Steamboats on the Western Rivers: An Economic and Technical History* (New York: Dover Publications, Inc., 1993), 6.
38 Ibid., 264–265.
39 David E. Schob, "Woodhawks & Cordwood: Steamboat Fuel on the Ohio and Mississippi Rivers 1820–1860," *Journal of Forest History* (July 1977): 124–125.
40 Hunter, *Steamboats on the Western Rivers*, 585.
41 "Lowell Factories," *Niles Weekly Register*, November 8, 1833 as quoted in Christopher F. Jones, *Routes of Power* (Cambridge, MA: Harvard University Press, 2016), 50.
42 Jones, *Routes of Power*, 6.

43 Hunter, *A History of Industrial Power in the United States Volume Two*, 400–401.
44 H.N. Eavenson, *The First Century and a Quarter of the American Coal Industry* (Baltimore, MD: Johns Hopkins University Press, 1942), table 20, 426–434. As quoted in Hunter, *A History of Industrial Power in the United States Volume Two*, 425.
45 George H. Thurston and W.S. Haven, "Pittsburgh as it is," *The Pittsburgh Quarterly Trade Circular* (1857): 24.
46 Jones, *Routes of Power*, 47.
47 Hunter, *A History of Industrial Power in the United States Volume Two*, 411.
48 Ibid., 427–428.
49 Joel A. Tarr and Karen Clay, "Perspectives on Coal and Natural Gas Transitions and the Environment," *in Energy Capitals: Local Impact, Global Influence*, eds. Joseph A. Pratt, Martin V. Melosi and Kathleen A. Brosnan (Pittsburgh: University of Pittsburgh Press, 2014), 6–7.
50 Timothy Mitchell, *Carbon Democracy: Political Power in the Age of Oil* (London: Verso, 2011), 19.
51 Walter Minchinton, "The Rise and Fall of the British Coal Industry," 223.
52 Ibid., 19–20. Mitchell's statistics are taken from P.K. Edwards, *Strikes in the United States, 1881–1974* (New York: St Martin's Press, 1981), 106.
53 Carter Goodrich, *The Miner's Freedom: A Study of the Working Life in a Changing Industry* (Boston: Marshall Jones Co., 1925), 19.
54 Mitchell, *Carbon Democracy*, 21–23.
55 Thomas G. Andrews, *Killing for Coal: America's Deadliest Labor War* (Cambridge, MA: Harvard University Press, 2008), 282–284.
56 Ibid., 288.
57 Lon Savage, *Thunder in the Mountains: The West Virginia Mine War, 1920–21* (Pittsburgh: University of Pittsburgh Press, 1990), xi.
58 Barry Supple, *The History of the British Coal Industry: 1913–1946: The Political Economy of Decline*, vol. 4 (Oxford: Clarendon Press, 1987), 6.
59 Ibid., 6–7.
60 Ibid., 163–164.
61 Ibid., 186.
62 Mitchell, *Carbon Democracy*, 37.
63 Daniel Yergin, *The Prize: The Epic Quest for Oil, Money, and Power* (New York: Simon & Schuster, 1991), 33. As cited in Mitchell, 36.
64 Lewis Mumford, *Technics and Civilization* (New York: Harcourt, Brace & Co., 1934), 235. As cited in Mitchell, 37.
65 Ibid., 272.
66 Ibid., 696.
67 Bruno Turnheim and Frank W. Geels, "Regime Destabilization as the Flipside of Energy Transitions: Lessons from the History of the British Coal Industry," *Energy Policy* 50 (June 9, 2012): 39–43.
68 David Severn, "England's Coal Legacy," *The New York Times* (June 28, 2015), SR 6.
69 www.sixtone.com/news/704/coal-story-china.
70 "World Coal Consumption Experienced a Record Drop in 2016," *Yale Environment 360*, (June 14, 2017), https://e360.yale.edu/digest/world-coal-production-experienced-a-record-drop-in-2016.
71 "Coal's Decline," *Union of Concerned Scientists* 17 (Fall 2017): 17.

References

Andrews, Thomas. *Killing for Coal: America's Deadliest Labor War*. Cambridge, MA: Harvard University Press, 2008.

Church, Roy. *The History of the British Coal Industry, Volume 3: 1830–1913: Victorian Pre-eminence*. Oxford: Clarendon Press, 1986.

Crosby, Alfred W. *Children of the Sun: A History of Humanity's Unappeasable Appetite for Energy*. New York: W.W. Norton & Co., 2006.

Eavenson, H.N. *The First Century and a Quarter of the American Coal Industry*. Baltimore, MD: Johns Hopkins University Press, 1942.

Edwards, P.K. *Strikes in the United States, 1881–1974*. New York: St Martin's Press, 1981.

Flinn, Michael, and David Stoker. *The History of the British Coal Industry, Volume 2, 1700–1830: The Industrial Revolution*. Oxford: Oxford University Press, 1984.

Geels, Frank W., and Bruno Turnheim. "Regime Destabilization as the Flipside of Energy Transitions: Lessons from the History of the British Coal Industry, 1913–1997." *Energy Policy* 50 (November 2012): 35–49. doi: 10.1016/j.enpol.2012.04.060.

Goodrich, Carter. *The Miner's Freedom: A Study of the Working Life in a Changing Industry*. Boston: Marshall Jones Co., 1925.

Hatcher, John. *The History of the British Coal Industry, vol. 1: Before 1700: Toward the Age of Coal*. Oxford: Clarendon Press, 1993.

Hatcher, John. "The Emergence of a Mineral- Based Energy Economy in England, c. 1550–1850." In *Economia e Energia, SECC.XIII-XVIII, Instituto Internazionale Di Storia Economica*, edited by Simonetta Cavaciocchi, 483–504. Firenze, 2002.

Hunt, E.H. *British Labour History*. London, Oxford University Press, 1981.

Hunter, Louis C. *A History of Industrial Power in the United States, Volume Two: Steam Power*. Charlottesville: University of Virginia Press, 1985.

Hunter, Louis C. *Steamboats on the Western Rivers: An Economic and Technical History*. New York: Dover Publications, Inc., 1993.

Jevons, William Stanley. *The Coal Question: An Inquiry concerning the Progress of the Nation, and the Probable Exhaustion of Our Coal Mines*. London: Macmillan and Co., 1906.

Jones, Christopher F. *Routes of Power: Energy and Modern America*. Cambridge, MA: Harvard University Press, 2016.

Lessing, Lawrence P. "Coal." *Scientific American* 193, no. 1 (July 1955): 58–67.

Mathias, Peter. *The First Industrial Nation: An Economic History of Britain 1700–1914*. London: Methuen & Co., 1983.

Mathias, Peter. "Economic Expansion, Energy Resources and Technical Change in the Eighteenth Century: A New Dynamic in Britain." In *Economia e Energia, SECC.XIII-XVIII, Instituto Internazionale Di Storia Economica*, edited by Simonetta Cavaciocchi. Firenze, 2002.

McElroy, Michael B. *Energy: Perspectives, Problems, & Prospects*. New York: Oxford University Press, 2010.

Minchinton, Walter. "The Rise and Fall of the British Coal Industry: A Review Article." *VSWG: Vierteljahrschrift Für Sozial- Und Wirtschaftsgeschichte* 77, no. 2 (1990): 212–226. www.jstor.org/stable/20735640.

Mitchell, Timothy. *Carbon Democracy: Political Power in the Age of Oil*. London: Verso, 2011.

Mumford, Lewis. *Technics and Civilization*. New York: Harcourt, Brace & Co., 1934.

Pollard, Michael. *The Hardest Work Under Heaven: The Life and Death of the British Coal Miner*. London: Hutchinson & Co., 1984.

Savage, Lon. *Thunder in the Mountains: The West Virginia Mine War, 1920–21*. Pittsburgh: University of Pittsburgh Press, 1990.

Schob, David E. "Woodhawks & Cordwood: Steamboat Fuel on the Ohio and Mississippi Rivers 1820–1860." *Journal of Forest History* 21, no. 3 (1977): 124–132. doi: 10.2307/3983286.

Seow, Victory. "The Coal Question in the Age of Carbon." Accessed November 4, 2018. https://sites.fas.harvard.edu/~histecon/energyhistory/seow.html.

Severn, David. "England's Coal Legacy." *The New York Times*, June 26, 2015, SR 6.

Supple, Barry. *The History of the British Coal Industry: 1913–1946: The Political Economy of Decline, Volume 4*. Oxford: Clarendon Press, 1987.

Tarr, Joel A., and Karen Clay. "Pittsburgh as an Energy Capital: Perspectives on Coal and Natural Gas Transitions and the Environment." In *Energy Capitals: Local Impact, Global Influence*, edited by Joseph A. Pratt, Martin V. Melosi, and Kathleen A. Brosnan, 5–29. Pittsburgh: University of Pittsburgh Press, 2014.

U.S. Bureau of the Census. *Historical Statistics of the United States, Colonial Times to 1700, Part One*. Washington, D.C.: U.S. Government Printing Office, 1975.

"World Coal Consumption Experienced a Record Drop in 2016." *Yale Environment* 360, June 14, 2017. https://e360.yale.edu/digest/world-coal-production-experienced-a-record-drop-in-2016.

Wu, Shellen Xiao. *Empires of Coal: Fueling China's Entry into the Modern World Order*. Stanford: Stanford University Press. 2015.

Yergin, Daniel. *The Prize: The Epic Quest for Oil, Money, and Power*. New York: Simon & Schuster, 1991.

4 Petroleum

"Liquid gold"

Introduction

Much like coal, we need to reach back in the geological history of the planet to describe petroleum's origins. Oil's primary producers consisted of small marine microscopic organisms 'of plant and animal matter. They lived for only a few days in Earth's warm waters. Its ancient shallow oceans were wonderfully suited for microscopic organisms. The process of living and dying repeated itself many millions of times over these many millions of years. A small fraction of these organisms became the source of energy from petroleum.

In the world of tectonic-plate uplift and volcanic eruptions and a world characterized by ice ages and periods of warming, the shallow oceans became a graveyard for these organisms. As rivers discharged sediment in the oceans over these millions of years, the weight from these repeated discharges, combined with the thermal temperatures from Earth's core, pressure-cooked the nutrient-rich microscopic life until it became a thick layer of fossilized organic carbon. This buildup became the elements for our future supply of liquid fossil fuel.

Chemical conditions transformed decaying microscopic organisms in the ancient oceans into petroleum. First, as estuaries deposit high concentrations of nitrates and phosphate nutrients from the ocean floor, similar nutrients from the land are deposited in the oceans. Second,

> [t]he key factor here is the accumulation of organic material in sediments should exceed the rate at which oxygen can diffuse from the overlying water to satisfy respiratory demands of the aerobic (oxygen breathing) bacteria that would otherwise be effective in consuming the incoming organic matter.[1]

Increasing pressure from the sediments and heat from Earth's core became the building blocks of petroleum. Temperatures ranging from 60 to 160 degrees Celsius (140–320 °F) transformed organic material into reservoirs of oil.

Microscopic organisms contain more hydrogen molecules than coal and can more easily be transformed under pressure into hydrocarbons. Crude oil is a combination of liquid and gaseous hydrocarbons. Once refined, the former becomes kerosene, gasoline and semisolid asphalt. Refined gaseous hydrocarbons become propane, butane, methane or natural gas. Gaseous hydrocarbons exist as billions of bubbles making the liquid crude oil percolate and press against the miles sediment that became solid sandstone. Some of it leaked through cracks and spilled onto the ground and released natural gas into the atmosphere.

Coal revisited

To distinguish coal's origins from that of petroleum, it is important to note that coal originated on land during the Carboniferous Period, 300 million years ago, with the birth and death of plant and animal matter in a tectonic world of tropical forests. Plants capturing solar energy grew and died mostly on the coasts of tropical and subtropical land. These coastal environments changed with the tides and sediments delivered to coastal margins from eroding continental rocks. A swampy and wet tropical landmass became the burial ground for ancient trees and giant ferns. Once an ice age replaced the warm interglacial period, dead plant matter became compressed under the expanding ice sheets. With additional cycles of warmth and cold over millions of years, new trees lived and died, sediment from continental erosion piled on. Tectonic processes fractured the supercontinents and began the movement that over millions of years positioned them in the alignment we know today. The location of the coal that we mine today is the product of tectonic movement and uplift. Chapter 3 provided a description of the different coal types.

The early history of petroleum

Since some oil leaks to the surface, its use in antiquity came from many sources. Cement using heated bitumen (thick, jellied crude oil) as mortar held fired bricks together in the construction of the walls and towers of Babylon as early as 2500 BCE. Pools of oil seeping from belowground continued to provide this Mesopotamian civilization with asphalt and pitch as mortar for the more elaborate walls built by Nebuchadnezzar II (634–562 BCE). The Mesopotamian civilizations of Sumeria, Babylonia and Assyria used bitumen mortar to seal gaps in the hulls of their boats and ships and as caulking for the wooden walls of their elaborate irrigation systems.

In the Caucasus region near modern-day Baku, major oil seeps provided the fuel for the "eternal flames" burning in the temples of ancient Persia about 1700 BCE. In the first century of the Common Era, the Roman natural philosopher Pliny the Elder (23–79 CE) wrote of the therapeutic properties of bitumen in treating toothaches, fevers, coughing and bleeding. Marco Polo wrote in his journals in 1271 CE that oil seeps near modern Baku:

> There is a spring from which gushes a stream of oil, in such abundance that a 1000 camels may load there at once. The oil is not good to eat; but it is good for burning and as a salve for men and camels affected with itch or scab. Men come from a long distance to fetch this oil, and in all the neighborhood no other oil is burnt but this.[2]

Other accounts describe earthen fissures of fire. We know such as escaping gaseous hydrocarbons ignited by sparks and lightning strikes.

With the death of Mohammad in 632 CE and the geographic expansion of Islam, extensive documentation about the uses of oil became available. Incendiary devices used oil in warfare. Muslims began distilling crude oil into kerosene for lamp oil as early as 850 CE. In addition to illumination in the ninth century, they mixed heated bitumen and water with sand to make asphalt for paving roads. In Europe, for comparison, the first asphalt-paved road appeared in Paris, France, in 1838 CE. No wonder that the modern global oil industry is centered in Southwest Asia, with Saudi Arabia, Iran, Iraq and the United Arab Emirates producing millions of barrels of oil daily.[3]

Ancient and medieval uses of oil rising to the surface were not restricted to Southwest Asia. California experienced similar oil seeps much before the Common Era. The Chumash people lived along the Santa Barbara channel region in 500 BCE. Millennia ago, their forebearers migrated across the frozen land bridge of the Bering Straits connecting Asia and North America. They used bitumen as an adhesive for attaching stone tools to wooden handles, to hold fibers together and as a coating for sewing string to fishing spears. Caulking for canoes allowed indigenous people from the region to travel farther and increase their trading networks. No wonder that modern California would become a significant contributor to the world's oil supply.

Ancient Chinese drilling technology

It was the Chinese, however, whose exploitation of subsurface minerals resembled modern drilling and the transportation of oil and natural gas. Evidence of oil and gas production in ancient China appears in ancient texts beginning in 60 BCE.[4] Drilling for brine as part of China's salt industry and its hydrocarbon industries remained separate activities until about 1500 CE. Drilling brine wells deeper, 300 to 400 meters (985–1313 ft) below the surface released natural gas deposits. Within decades, methods for capturing natural gas and burning it under big pans burned off the liquid brine and left the pans filled with salt. Bringing brine and natural gas to the surface industrialized Chinese salt production for human consumption and as a preservative for food. In Sichuan Province alone, sinking wells to depths of 700 to 800 meters (2297–2625 ft) resulted in the production of 150,000 tons of salt in 1850 CE for provincial consumption and for a growing salt trade.

Sometime during the beginning of the Common Era, Chinese well diggers stopped using shovels to unearth brine deposits and invented percussion drilling, with a drill bit made of iron and a pipe of locally harvested bamboo. Building a rig of bamboo with men standing on a seesaw lifted the pipe a meter or more. By using the weight of the men, the pipe crashed into the surface, with the iron bit smashing the rock below. Blow by blow, using the energy of humans, this technology drilled wells up to 140 meters (460 ft) by 400 CE. Surprisingly, with its variety of bits and methods to repair collapsed shafts, the tools and techniques for drilling mimicked modern drilling technology.

The origins of the commercial oil industry in the United States

In 1852, Francis Beattie Brewer moved from Vermont to Titusville, Pa. to join a lumber business on Oil Creek owned partly by his father. On the company's property, oil seeps discharged crude oil onto the surface where it was collected to lubricate company machinery. Residents bottled it, claiming that it possessed medicinal qualities for treating eruptions on the skin, burns, cuts and bruises. Distilled and swallowed, it allegedly cured some diseases.

Returning to New England the following year, the younger Brewer carried a bottle of "Creak Oil" to share with his uncle, Dixi Crosby, a professor of surgery, and Oliver Hubbard, a professor of chemistry at Dartmouth Medical School. The events that followed culminated in a trip by Crosby's son, along with the younger Brewer, back to Titusville. Both men visited the oil springs located near Oil Creek, the future site of Oil City. Brewer remembered, "We stood on the circle of rough logs surrounding one spring and saw the oil bubbling up, and spreading its bright and golden colors over the surface, it seemed like a golden vision."[5]

To satisfy potential investors analyzing the chemical properties of oil, professor of chemistry Benjamin Silliman Jr. wrote, *Report on the Rock Oil, or Petroleum, from Venango Co.* (1855). It concluded that "your Company [Pennsylvania Rock Oil Co.] has in its possession a raw material from which, by simple and not expensive process [fractional distillation] may manufacture very valuable products."[6] One of these products was distilled lamp oil that burned brighter and lasted longer than candles. By comparison, whale oil was too expensive, and camphene was too explosive. Manufactured coal gas (a topic of the next chapter) required the capital investment in infrastructure and piping. In an increasingly industrialized and urban world of the United States and Europe, better and cheaper artificial lighting became a prerequisite for modern living. Textile mills with small windows and cramped urban living quarters required more than the natural lighting provided by the sun. New machinery, with its complex gearing and multiple moving parts, needed new solvents to lubricate them more effectively than older crude oils.

A visit to Titusville by two of the company's investors confirmed the presence of oil in different locations. They reported that, "digging down into the hardpan, the oil mixed with water seemed to rise up as though there was pressure beneath. The deeper the excavations, the more abundant the oil appeared."[7] They did not know, nor could they have known that the Seneca, First Nation inhabitants of the area, long before the arrival of European settlers in the eighteenth century, gathered small amounts of oil from seeps for decorative purposes. By 1755 CE, settlers to the area noted the presence of oily waters. So, the findings of these two investors simply confirmed what inhabitants to the Oil Creek Valley had known for the past millennia.

If timing for an investment is "everything," then 1857 was a poor time to seek capital from a newly formed stock company. The financial Panic of 1857 sent Wall Street investors scrabbling. Given the movement of the investors and stakeholders out of the company at the time of the financial panic, two stakeholders, George Bissell, a New York attorney, and James Townsend, president of a New Haven, Connecticut, bank, remained key figures in the company's efforts to begin drilling for oil in the Titusville/Oil Creek region.

Living in a New Haven hotel in the summer of 1857, Townsend became friends with Edwin L. Drake, a retired railroad conductor with the now-bankrupt New York and New Haven Railroad. Recently widowed with children and looking for a way to earn a living to support his family, Drake also lived in the hotel. Possessing few technical skills but showing great interest in a petroleum "start-up" company that Bissell and Townsend led, Drake, armed with a free railroad pass, agreed to travel to Erie, Pennsylvania, and ride the rough forty miles on a stagecoach to the tiny lumber town of Titusville. For his efforts, Townsend leased the previously purchased Hibbard farm to Drake with a 12 cent per gallon royalty rate. At the same time, Townsend dealt with rebellious investors by organizing a new company, Seneca Oil Company, in March 1858.

A month later, Drake, looking for a driller to begin operations, traveled back to Titusville. A year passed before he found a driller named William Smith and enough workers to construct a primitive wooden derrick, build a machine shop and assemble a steam boiler. Smith proved to be a skilled craftsman capable of repairing machinery, but when he could not, Drake was sent packing to locate replacement parts. The drilling itself presented numerous problems, not the least of which was cave-ins and groundwater flooding the well. On August 28, 1859, oil mixed with water rose to the surface of the well after the drill bit had cut 71 feet into the ground and slipped another 6 inches into a crevice containing a pool of crude. What Smith saw was "a dirty greenish grease."[8]

From such an inauspicious beginning began an oil boom that changed the nation's fortunes, and as historian Brian Black has noted, "[w]hat had once been valueless now was a commodity of skyrocketing worth-locally, regionally, and nationally. In this valley, invisible and beneath the surface of the earth, lay tremendous natural-and indeed national-treasure."[9] A boom of wildcatters rushing to Oil Creek followed Smith's discovery that warm summer morning.

Derricks appeared quickly, scattered in the valleys, on hillsides and hilltops. Those that produced oil remained; those that drilled dry holes were abandoned, littering the landscape with derricks and drilling paraphernalia. The rule of capture meant that tapping a pool of crude that extended beyond the boundaries of the leased land belonged to the first one to strike oil and bring it to the surface. The rush for oil brought all sorts of persons, ranging from unemployed workers to land speculators, to Oil Creek seeking "get rich fast" schemes. By 1864, the effects of this uncontrolled exploitation of this newfound mineral wealth were visible for all to see.

> The creek, its flat banks covered with derricks, vats and engine houses jostle each other in confusion. The creak and wheeze of the engine and the

Figure 4.1 Early wooden oil derricks in Titusville, Pennsylvania.

> pump are mingled with the shouts of the flatboatsman urging his team of four horses abreast up the middle of the stream with a load of barrels.[10]

Humans, horses, woodlands and waterways combined with coal-burning steam engines to bring liquid and gaseous fossil fuel to the surface. The flow of organic energy in all its forms combined to promote the transition to a mineral energy regime in the industrializing world.

Derricks producing oil appeared almost everywhere in the Oil Creek valley. Farmers sold or leased their land for dollars an acre in 1859. A few years later, they saw prices rise into the hundreds of thousands per acre. The first derricks produced dozens of barrels of oil in 1858. Two years later, the derricks along the 3- to 4-mile stretch of the valley produced 450,000 barrels. By 1862, more than 3 million barrels gushed from the Oil Creek wells.[11]

Marketing petroleum

Manual laborers collected percolating oil from the wells by digging pits surrounding the well. Later, wooden vats would become containers for crude. Then, they gathered it into barrels for transport by wagons pulled by teams of horses to collection points along rivers and roadways. Human labor, horsepower and wood converged to accelerate the transition to a mineral energy regime. Getting the crude to market posed another problem. Transportation costs exceeded those of drilling by a large margin. Getting crude oil out of the ground represented the first stage in marketing the product. Barrels, wagons, roads and water transport on barges preceded the building of pipelines.

In Titusville, Oil Creek became the logical highway for carrying oil to its next destination on the Allegheny River, 16 miles below Titusville. Since the shallow depth of the Creek served as a natural barrier, lumbermen, employed by oilmen, built a series of temporary dams upstream to raise the water level. Behind the farthest dam, loaded oil barges waited for the moment when water levels reached a depth of 2 feet. In unison, workers removed the wooden dams, allowing the barges carrying thousands of barrels (a barrel is 42 gallons) to float downstream. "A December 1862 disaster triggered by breaking ice destroyed 350 boats and 60,000 barrels of oil and caused $350,000 of property damage."[12] Clearly, the unpredictability of water transport and the dangers encountered constrained the expansion of the industry into growing urban markets.

With few paved roads, spring floods made many impassible. When they were passable, teamsters driving their horses with loads of barrels clogged the roads on their way to the nearest rail depot. With a monopoly on short-haul transport, teamsters' costs exceeded the railroad cost from the depot to cities as far away as New York. Drilling and barreling oil with many fewer laborers cost less than the mining and transport of coal. By eliminating the monopoly role of teamsters, "[o]il pipelines were invented as a means of reducing the ability of humans to interrupt the flow of energy."[13]

Conditions in coal mines promoted worker solidarity. In the mines, large concentrations of workers in dangerous, confined quarters engaged in exhausting manual labor. Solidarity led to expressions of shared grievances, demands for higher wages, safer working conditions and unionization. Examples of strikes against mine operators and violent confrontations that led to injuries and deaths became common occurrences in Britain and in the United States. Conditions in the oil fields differed in that considerably less human labor was required to drill, transport, refine and market oil products. A smaller workforce in oil produced a considerably larger quantity of energy than that in coal. Strikes in the oil fields happened but did not define worker/management relationships in ways similar to mining.

In oil, workers remained aboveground working in proximity to managers. With oil transported by pipelines and pressurized by steam-powered pumps, rather than wagons or railroads, fewer workers engaged in loading and unloading the fuel. Moving a lighter liquid fuel by pipelines instead of heavy loads of coal by railroad simply required fewer handlers, so fewer opportunities existed to establish worker solidarity to improve working conditions.[14] A barrel of oil provided one-quarter of the energy produced by burning a ton of coal. "Pipelines produced a landscape of intensification that furthered the dramatic growth in petroleum consumption and the deepening of the mineral energy regime."[15]

Once installed in 1863, wooden pipelines circumvented the wage demands of teamsters and eliminated the unpredictability of weather conditions on the roads. Within two years, Samuel van Syckel, a producer, built a 2-inch-diameter wooden pipeline powered by three pumps driven by steam engines that connected oil wells to the region's railroads. "Astute oilmen learned a valuable lesson: rather than negotiating or dealing with laborers and sharing the industries profits, it was easier to replace them with technology powered by mineral energy."[16] While Oil Creek remained the center of early drilling, oil fields in Ohio, western Virginia and the present-day location of Ontario, Canada, began producing oil. Surpluses would continue to mount and prices would continue their slide unless refined crude oil could compete with manufactured coal oil manufactured by more than 60 firms by 1860 as an illuminant and lubricant. The transition from one to the other, albeit complicated occurred swiftly.

Production outran capture and storage capacity at many of the wellheads. Liquid and gaseous hydrocarbons comingled in their pressured packed subterranean space. In 1861, at a well owned by H.R. Rouse, one of the first gushers capable of producing 3,000 barrels of crude a day, exploded. Drilling into that space spontaneously released natural gas rushing up the well and ignited by a spark from steam engine nearby. Crude oil followed with the explosive force of a hurricane. This spontaneous and uncontrolled gusher spewed massive amounts of burning crude 60 feet into the air, soaking the surrounding area with blazing oil. Fed by escaping flammable gas, the explosion incinerated everything, including Rouse and nineteen workers. Five days passed before workers extinguished the fire. Recognizing the highly flammable nature of the gas led operators to ban smoking near wellheads.

The damage done to the formerly pristine land around the wellheads was complete. The repeated saturating of the surface with crude oil transformed the surrounding environment. Oil seeped into the creek and flooded the land. Digging pits to contain and collect oil resembled small ponds:

> The soil is black with waste petroleum. The engine-houses, pumps, and tanks are black, with the smoke and soot of the coal-fires that raise the steam to drive the wells. The shanties – for there is scarcely a house in the whole seven miles of oil territory along the creek – are black. The men that work among the barrels, machinery, tanks, and teams of white men blackened . . . Even the trees wore the universal sooty covering. Their very leaves were black.[17]

To exacerbate the environmental conditions caused by waste petroleum, coal-fired steam engines, used to power the drilling and pumping machinery, filled the air with toxic fumes and soot composed of ash and tar. Such a volatile mix made living and working conditions a toxic brew. The frenetic activity of men and machines, of derricks, flowing and barreled oil, steam engines and barges signaled progress. At the same time, one experienced a degraded environment. Its denuded land stripped of its trees was turned into lumber to build oil derricks. The constant threat of fire from exploding natural gas put investors, operators and workers on high alert.

Flammability and oil wells were an order of magnitude much larger than those experienced by mining coal. The combination of gaseous and liquid hydrocarbons released in close contact with coal-fired machinery caused numerous explosions. Workers transporting and using nitroglycerin "torpedoes" to fracture subterranean rock formations added volatility to an already dangerous environment. In addition, negligence about potentially lethal combinations played a role.

Mined coal required a conscious intervention for ignition. Fires occurred in mining coal from a combination of poor ventilation, poor visibility, partially compensated by illuminants, namely, candles, oil-wick and carbide lamps. Acetylene gas powered the latter and burned brighter, but was capable of igniting the methane gas found in coal mines, causing explosive fires. Safely, lamps with the flame enclosed behind a mesh covering proved to be much safer. Not until the invention of the Edison Flameless Electric Cap Lamp powered by a battery pack the early twentieth century did mine fires diminish.

As the number of oil wells grew throughout the United States and Canada, the number of steam engines put in place to pump crude out of the ground grew accordingly. Initially wood powered these engines because of the proximity of forests to the wells. As timber cutting led to deforestation and eliminated its closeness to wellheads, transportation costs for barrels of crude on horse-powered wagons and railroads powered by burning wood became prohibitive. Coal with more energy by volume than bulky cords of cut timber began to power oil-drilling machinery.

The market for illuminants and lubricants: organic oils and mineral oils

By 1864, manufacturers had accelerated the transition from coal oil, which along with plant and animal oils had dominated the market for illuminants and lubricants. Many factors caused the relatively speedy transition. Manufacturing costs differentials led the way. Coal-oil production required the use of a retort to burn coal at about 850 degrees Fahrenheit, allowing the oily vapors to escape from the neck of the retort into a condensing chamber. In this way, the retort distilled oil from coal.

Kerosene dominated the market for illuminants and lubricants. The Canadian geologist Abraham Gesner had distilled coal into coal oil and received a patent, naming it by the trade name *Kerosene*. In 1850, he founded the Kerosene Gaslight Company and began the process of replacing animal oils, including the more expensive whale oil, with his patented product, a combustible hydrocarbon. Manufacturers of coal oil for lighting burned an estimated 60,000 bushels of coal (a bushel weighs 80 lb) daily to produce 75,000 gallons of crude coal oil.[18] With Drake's well producing about 400 gallons of oil a day, the volume reached 10,000 gallons stored on-site in wooden vats in no time at all. Competing with coal oil represented the first problem. Fortunately, he was able to sell it cheaply at 25 cents a gallon as lamp oil and refined as a lubricant for machinery. Overproduction during the 1860s led to a collapse in prices, to as low as US$2.40 a barrel, and caused many drillers to leave the field.

Refining crude petroleum required different processes for the different grades of illuminants and lubricants. Crude oil is heated in a furnace and separated into fractions by distillation. The fractions at the top boil at lower temperatures (20–150 °C) and are distilled into propane and petroleum. The next fraction distills kerosene (200 °C). The next fraction is diesel fuel (300 °C). The fractions at the bottom, including gasoline, lubricating oil, paraffin and asphalt are heated at 370 to 400 degrees Celsius. These are lighter products. Further processing into fractions results in more products.[19]

Standard oil of Ohio

The differences in distilling coal oil from coal and the number of products distilled from crude petroleum made economic sense. Once the differences became apparent to coal oil producers, they embraced the benefits and converted their factories using retorts for coal oil to refineries for petroleum-based products. Petroleum refineries co-opted the patented name for Gesner's product and began using kerosene (small *k*) to identify all mineral oils. In this way, petroleum replaced coal oil.[20] "[John D.] Rockefeller formed the Standard Oil Company of Ohio in 1870; by 1879 Standard controlled 90 percent of the U.S. refining capacity, most of the rail lines between urban centers in the Northeast, and many of the leasing companies at the various sites of oil speculation."[21] How did Rockefeller accomplish so much in such a short time? Initially, the use of the word *standard* in the company's title was intended to promote the idea that Rockefeller's products guaranteed the highest standard in the refining process. By implication, his competitors' products lacked uniformity and high standards.

The importance of symbols in marketing products notwithstanding, Rockefeller's ruthless determination to drive his competitors either into bankruptcy or "into a corner" from which they could do little but accept his demands knew no bounds. At the age of 16, in Cleveland, he took a job with a firm selling vegetables. In 1859, he formed a partnership with Maurice Clark and did so again during the Civil War, opening of the Great West to migration. Cleveland benefited from the oil boom in Oil Creek with the building of a railroad linking the oil fields to refineries into Cleveland. The partners entered the oil refining business and as it grew Rockefeller bought out Clark. By the end of the Civil War in 1865, he owned the city's largest refinery and was wealthy enough to expand his business.

In 1866, he built a second refinery and started an export firm in New York City with his brother William in charge to locate markets for the refineries growing capacity. Sales reached US$2 million that year while Rockefeller bought oil from drillers, bought land to grow white oaks to make barrels and later bought tank cars to replace barrels and sell kerosene by the gallon to households from horse-drawn tanks. Getting crude oil to refineries on railcars was the greatest single variable cost. By putting all profits back into his business and borrowing money, he created a firm in 1870 and named it Standard Oil of Ohio; it was already larger and more profitable than his competitors. Rebates on guaranteed large and regular shipments of crude and refined oil became the vehicle for demanding and receiving rebates from rail companies that undercut the costs charged to his competitors.

Getting rebates on shipping costs encouraged Rockefeller to demand even more. As Standard Oil became more profitable, adding refinery capacity with its newfound wealth, it demanded and received "drawbacks." As oil expert Daniel Yergin has pointed out, a competing refinery would pay, for example, one dollar a barrel to ship its oil to New York. The railroad paid US$0.25 back to its competitor, Standard Oil. Already paying less for transportation through rebates than its competitors, such rebates subsidized Standard Oil.[22] By 1870, it refined so much more oil than its competitors that they dominated the shipping industry with more and more large guaranteed shipments. Posted prices per barrel ranged from US$1.25 to US$1.40, but Standard Oil paid US$0.85 or less.[23] During the 1870s, he either bought out his competitors or undercut prices for kerosene and lubricating oil, crude oil's largest and most profitable refined products, that forced them to quit the refinery business.

Kerosene

The use of kerosene to light the darkness transformed the lives of many millions of people throughout the world. For the majority, daylight and the onset of nighttime defined human activity. Sunlight defined most human activities, and when it disappeared, humans retired for the day, awakening with the rising sun the next day. Kerosene, an illuminator, lengthened the day for the first time in human history. Granted that candlelight, gaslight, firelight and camphene (strong odor) served as illuminators before refined kerosene was available, some emitted soot; others were costly; some increased the threat of fire, while others cast a weak glow.

Lamps filled with whale oil, coal oil or camphene had existed for decades. Consumers knew the drill of filling a lamp's bowl with oil and lighting the wick that drew oil from the bowl and had learned to trim the wick as it degraded from repeated use. By the 1850s, a dollar bought most functional lamps, so they became available to most households. Replacing the other oils with cheap, clean-burning kerosene that cast a strong, steady glow required no capital costs. In 1863, a gallon of kerosene cost US$0.50 wholesale and dropped to US$0.25 by 1871 in New York.

Cheap kerosene was superior in its glow to other illuminants. "Analyzing the cost of the light output of various illuminants in America in 1870, economist William Nordhaus found that kerosene had a seven-to-one advantage over camphene, a twelve-to-one advantage over manufactured gas, and a twenty-to-one advantage over sperm whale oil."[24] Internationally, kerosene became the illuminant of choice as safely standards improved and prices continued a steady decline. In Britain, a million hours of light cost a low of 200 pounds in 1900, and the country's consumption rose from 3.3 billion hours in 1870 to 1.5 trillion by 1900. Kerosene penetrated foreign markets quickly overcoming the shortage of fats and oil that had plagued Europe for a generation.

Exports of kerosene accounted for more than half of the United States oil exports in the 1870s and 1880s. Its value was the country's fourth largest, behind cotton, wheat and meat, and the first among manufactured goods. One could claim that it came from one country, the United States, but in fact, it came mostly from one locale, Oil Creek, Pennsylvania.[25] Pennsylvania also provided the coal to power the steam engines that drilled the oil wells in Titusville and powered the pumps that delivered the crude to rail depots through its pipelines. The ubiquitous noise of steam engines symbolized progress. As one form of mineral energy became instrumental in the development of crude oil extraction and refining.

Quality and price catapulted kerosene to a dominant position as an illuminant at home in the United States and internationally. The agent that quickly came to control its global reach would be none other than Standard Oil. By the end of the 1870s, it controlled 90 percent of the exported product, having refined it and through its network of railroad contracts delivered it to American ports for delivery internationally.

National and international competition

Despite Standard Oil's early dominance, events elsewhere, particularly in Czarist Russia, would result in the development of a competing oil industry. The oilfields of Baku, in modern Azerbaijan, had been known to residents and to travelers beginning with Marco Polo in the thirteenth century. He learned from residents whose hand-dug pits produced oil, good for burning and cleaning the mange off camels but nothing else. Annexed by Russia in 1806 CE, a fledging oil industry existed with 82 hand-dug pits producing small quantities. Decades later, steam-powered percussion drills replaced the hand digging of wells using human and animal labor.

A surge in production at Baku followed as mechanization allowed drillers to penetrate deeper layers of sandstone to release oil. A surge followed making Baku an oil capital for a short time in the early twentieth century when it produced more than one-half of the world's total supply of oil. Its heavy crude, however, yielded less refined kerosene and a higher volume of residual oil for steam boilers. In this sense, it did not compete with Standard Oil for illuminants. In that market, Rockefeller controlled 80 percent of the sales by cutting prices and acquiring potential competitors and through innovation. He replaced cumbersome, leaky and expensive barrels with railway tank cars for long hauls and horse-drawn tank wagons for local sales.

The Baku oil fields flourished with an infusion of money and managerial skill from the Nobel Brothers Petroleum Producing Company. Its oil fields produced 10.8 million barrels in 1884 CE with nearly 200 refineries working in a new industrial suburb of Baku known as "Black Town" because of its foul oil smoke and its polluted landscape. Producing one-half of Russia's kerosene, the Nobel Brothers Company drove Standard Oil out of the country's oil market, despite its heavy crude limiting its volume of refined kerosene. Expanding its market beyond the Russian Empire into its European neighbor countries required the building of a railroad from Baku to the port city of Fiume on the Adriatic Sea. There, the

Rothschild family of France, who owned an oil refinery in Fiume, intervened. Having financed the construction of railroads throughout Europe, the Rothschilds financed the completion of the railroad in exchange for leases on Russian oil fields and guaranteed shipments to Europe at below-market prices. Production rose to 23 million barrels between 1879 and 1888 CE, reaching more than four-fifths of American production. Standard Oil's share of the international kerosene market fell to 71 percent in 1891 while the Russian share rose to 29 percent.[26]

With the international oil market controlled by Standard Oil and the Nobel/Rothschild Companies in Russia, the world was afloat in refined oil products. Both competitors looked to Asia as opportunities to expand its share of the global market. In the process, they discovered new crude oil reserves. The Rothschilds entered the quest for new global markets, first, by enlisting the help of the Marcus Samuel family, British merchants, whose initial business included the purchase and sale of seashells as curios to London's girls and women.

Entering the global oil competition, the Samuels made their fortune building and leasing newly designed oceangoing oil tankers and building large storage tanks in Asia's many ports to receive and store refined kerosene. The Samuels had laid the origins of what would become Shell Transport and Trading Company in 1897. By the 1870s, global communication shortened the geographical distances between the West and the East. The Suez Canal opened to oceangoing traffic in 1869, eliminating 4,000 miles of travel between both regions. A year later, a direct telegraph cable linked London to Bombay, India, followed shortly thereafter by connections to Nagasaki, Japan; Shanghai, China; Singapore; and Sydney, Australia.[27]

In Asia, new oil discoveries in the Dutch East Indies island of Sumatra represented the world's third-largest oil-producing region. It quickly became a major supplier of kerosene in the Asian market. With powerful sponsors in the Netherlands, the country's king, William III, recognized the company's potential growth by granting the right to change its name from Crown Oil to Royal Dutch Petroleum in 1890.

Before the 1880s, Oil Creek represented the only significant oil-producing region in the world. Also, Standard Oil's control of refineries and pipelines that turned Pennsylvania crude primarily into kerosene-dominated sales in the United States and across the world. By 1900 CE European companies had developed large oil production capacity in Baku and Sumatra. With advances in technology, this liquid energy source required fewer workers than coal mining and transported oil by "steel pipelines, high pressure steam pumps, bulk tankers and large storage tanks."[28]

Efforts by Standard Oil failed to either intimidate its competitors by slashing prices or buy out its rivals. Cooperation seemed to be the only viable alternative. So, during the early decades of the twentieth century, it and its major European competitors cooperated by dividing up a growing and profitable global market for kerosene. In Europe, two alliances gained control of the European and Asian markets. In 1902, Royal Dutch and Shell with financial support from the Rothschild family formed Royal Dutch/Shell. Three years later Shell's renewed might forced Burmah Oil in Rangoon, which controlled the growing Indian market for illuminants, to share its bounty with Shell and Standard Oil. A year later, in 1906, Rothschild joined Nobel, with financial backing from Germany's Deutsche Bank, to form the European Petroleum Union, a cartel controlling 20 percent of the European market for kerosene and fuel oil, agreeing to Standard Oil's claim for the majority share of 80 percent.[29]

California and Texas crude

As the world's productive capacity to drill for oil expanded, new oil fields in the United States replaced the dominant role of Oil Creek in Pennsylvania. Discoveries in California shifted the focus from the east to the west coast of the United States. By 1910, the state produced 22 percent of the world's production, or 73 million barrels, more than any foreign country. As historian Paul Sabin has pointed out, the sparsely populated San Joaquin Valley, densely populated Los Angeles and Southern California's coastal waters contained most of the state's oil. Technological advances in the early decades of the twentieth century increased production.

Oil companies, especially Union Oil and Standard Oil of California, learned not only how to drill deeper but to drill in a slanted direction. They also learned to extract more energy from a barrel of refined crude. Dozens upon dozens of wells produced so much oil that storage tanks flowed over and prices collapsed. How to promote demand and eliminate the wild fluctuation in prices meant that the California legislature needed to both stimulate oil production but restrain it by gas tax policies. Such policies funded highway construction and maintenance for vehicles powered by internal combustion engines burning gasoline refined from California crude oil. They promoted tourism by making its beaches and parks, recipients of tax dollars, accessible to residents and travelers.[30] Similar policies both at the state and federal levels would use the California model to fund highway construction and maintenance to the detriment of railways and mass transit development.

Before a federal highway construction program emerged as a national defense priority in the 1950s, California refined crude would find a market through Standard Oil in Asia, not the eastern and more heavily populated part of the United States. The proximity of Texas oil exploration and development would come to satisfy eastern demands for oil. Dismissed by geologists and Standard Oil, only a few Texans believed that an oil bonanza awaited wildcatters willing to drill in an area where gas bubbles percolated on a hill in the little town of Beaumont. Faced by a disbelieving majority of investors, the Pittsburgh wildcatters James Guffey and John Galey accepted the challenge and assumed the risk by insisting on a majority share of potential profits.

Using the new technology of rotary drilling and reaching depths exceeding 800 feet through sand that had frustrated previous wildcatters, an unpredictable explosion happened on January 10, 1901:

> Gas started to flow out; and then oil, green and heavy, shot up with ever-increasing force, sending rocks hundreds of feet into the air. It pushed up in

> an ever-more-powerful stream, twice the height of the derrick itself, before cresting and falling back to the earth.[31]

The Texas oil boom commenced with news of this gusher. Projections that it would produce 50 gallons of crude a day missed the mark by a wide margin. Lucas 1, named after Anthony Lucas, the engineer who pioneered the drilling expedition on Spindletop in Beaumont, Texas, produced 17 million barrels a year. Within months, fortune hunters descended on the area and drilled as many as 214 wells on the hill. The town's population swelled from 10,000 to 50,000, with as many as 16,000 living on the hill in tents. Crime, vice and mayhem ruled the area. Mud, oil slick and air pollution from the steam and internal combustion engines powering the drilling operations permeated the landscape. Deafening noise from machines and people became a constant intrusion. "According to one estimate, Beaumont drank half of all the whiskey consumed in Texas in those early months. Fighting became a favorite pastime."[32]

Unlike other oil fields, the chemical composition of Texas crude made it suitable for fuel oil to power machinery and locomotion but not as an illuminant. Noticed by Marcus Samuel, principal owner of Shell Transport and Trading Company, he wanted to convert his oceangoing fleet of ships from coal burning to oil. An arrangement with James Guffey to buy one-half of his annual production, a minimum of 15 million barrels at 25 cents a barrel would also allow him to diversify away from dependence on Russian oil. In the decade that followed oil strikes similar to the one that gushed forth at Spindletop, new ones in the Gulf Coast and in Oklahoma replaced the early advantage of Texas, with the Mellon family of Pittsburgh purchasing Guffey's stake in Spindletop and expanding it into what would eventually become Gulf Oil. With the separate emergence of Texaco Oil, Texas regained its primacy as the country's leading producer of crude oil in 1928.

Middle Eastern oil

With oil production accelerating around the world, controlling supply to prevent prices from collapsing became a priority for Standard Oil and its growing number of competitors. Shell Oil learned quickly that its guaranteed price of 25 cents per barrel for Texas crude proved to be no bargain when market prices fell to 3 to 4 cents a barrel as oil storage tanks swelled with unsold supplies. More discoveries of oil would exacerbate the existing glut. The large companies profiled in these pages knew that the Middle East contained many potential oil sites. Developing a new large oil field would disrupt the unstable equilibrium that the major companies sought to maintain. Also, Russia's Baku fields, among the world's largest producers, failed to gain a significant place in the global market for kerosene and fuel oil. If Russia built a pipeline to the Persian Gulf for the purpose of entering the Asian market to compete with Standard Oil, Royal Dutch/Shell and the others, more oil would further depress prices.

Fearing either, or both, potential developments, three of Europe's major oil firms responded by buying the rights to explore for oil in the Middle East from the region's Ottoman's rulers. The new European Petroleum Union explored the most promising site in Mesopotamia (modern Iraq, especially Mosul and Baghdad). Burmah Oil, founded in Scotland in 1886, became Anglo-Persian Oil with its explorations in Persia (modern Iran). Royal Dutch/Shell entered Egypt for exploratory activity. Using Standard Oil's recipe for controlling distribution and transportation by railway or pipeline, the European firms signed exclusive and proprietary agreements with the Ottoman rulers to prevent access from other oil companies. By controlling access to hubs and ports, Europeans placed constraints on the development of new oil fields in the Middle East and thereby limited global supplies and stabilized prices.[33]

European governments and the United States used other strategies to limit the development and production of Middle Eastern oil. They included the tacit agreement among the truly big international companies, including Standard Oil, Royal Dutch/Shell and the European Petroleum Union, to control worldwide distribution and marketing and government quotas and price controls in the United States. Despite efforts to curtail the exploration of oil in the Middle East by the methods outlines earlier and by agreements to delay the construction or completion of railways and pipelines to port cities in the Persian Gulf and elsewhere, drillers employed by Burmah Oil discovered a large oil field at Masjid-i-Suleiman, Persia, in May 1908. Once again, delay became the order of the day.

It took the company three years to lay an 8-inch pipeline the 140 miles to the border of Mesopotamia (Iraq) and to build refineries. The discovery of large oil fields in Mexico from 1910 to 1914 added to the potential glut and made it the third-largest oil-producing country in the world at that time. A British firm, Mexican Eagle, controlled 60 percent of that production and signed a supply agreement with the British government's Admiralty in July 1913, 13 months before the outbreak of the Great War (World War 1, 1914–1918). Standard Oil tried to undermine Mexican Eagle by supporting the overthrow Mexico's government led by President Porfirio Diaz. Revolutionary forces, some led by Emiliano Zapata, got control of the oilfields, thwarting arrangements with Britain.[34] Mexico nationalized its oil production in 1938 by setting up a national company, Petroleos Mexicanos (Pemex), eliminating British and U.S. control. To compensate for the loss, both had invested in oil production in Venezuela, replacing Mexico as the world's third-largest producer in the 1930s, behind the United States and the Soviet Union.

Oil's geopolitical impacts

"The availability of inexpensive oil encouraged the United States to adopt patterns of socioeconomic organization premised on high levels of oil use."[35] When domestic supply no longer supported the growing demand for oil, as happened when imports climbed to 50 percent for the first time in 1998 and peaked at 60 percent in 2005, U.S. national security interests coincided with control of foreign supplies of oil. Cuts in foreign supplies, mostly from the Middle East, shocked the mobile American "way of life," as we will see in the pages that follow.

The history of oil, as we know it, does not begin with the end of World War I but with understanding oil's unique strategic importance in national affairs and stems from decisions made between 1918 and 1920.

To place the role of large oil firms, the United States and Britain in historical context, the United States began converting its naval fleet from coal to oil shortly before World War I. Britain followed by converting to oil. It transformed its coal-burning battle fleet powered by its substantial coal reserves, its active coal-mining industry and coaling stations around the world. The policy had both positive and potentially negative consequences. On the positive side, converting from coal-fired steam boilers to steam turbines to drive the new electric generators powered by oil meant that human labor would be curtailed since many fewer men would be needed as shovelers and stokers. Coal filled multiple bunkers, while oil required less storage space. As a fluid, oil would flow mechanically to heat boilers. With more energy per volume, oil enables ships to travel further and faster with less cumbersome refueling.

In Britain, the substitution had potentially negative consequences. The substitution of oil for coal would deprive Britain of energy self-sufficiency and make the country dependent on imports from big oil firms. Ensuring one's national security by gaining and maintaining access to oilfields would become one of the government's primary objectives. Once Winston Churchill, then in charge of the Admiralty, committed it to oil in 1913, "he forever compromised the nation's energy autonomy with Anglo-Persian Petroleum becoming the most sensible option to ensure Britain's energy future."[36]

During World War 1, oil consumption grew by 50 percent. The United States produced 1 million barrels annually, or about two-thirds of the world's output by 1920. As the world's leading oil producer for three-quarters of the twentieth century, the United States faced no potential downside. While coal-fired factories produced the munitions and armaments for war, gasoline-powered motor vehicles – tanks, trucks and cars – drove men and machines to the battlefront. The role of ships, submarines and airplanes powered by fuel oil further highlighted the importance of highly mobile machines in winning the Great War (World War 1). Germany's defeat eliminated it from a position of strength among modern nations.

The Bolshevik Revolution in Russia in 1917–1918 (the Soviet Union in 1922) led to the nationalization of its oil fields. The global supply of oil remained in the hands of private corporations located in a comparatively small number of places, including the United States, Mexico, Venezuela, the Middle East and Indonesia. With nationalization in 1938, Mexico removed its oil from the control of giant oil corporations in Britain and the United States. The use of oil continued unabated following the war. Chemical warfare led to postwar commercial applications as petroleum-based pesticides, herbicides and fertilizers that initiated the Second Green Revolution in agriculture. Americans entered the automobile age powered by the gasoline-guzzling internal combustion engine, with registrations leaping from 3.4 million in 1916 to 23.1 million by 1930 and continuing their upward climb into the twenty-first century.

Unlike Britain, the United States and the Soviet Union possessed large oil national reserves. In the United States, discoveries in Texas, Louisiana, Oklahoma, California and Alaska increased the country's energy consumption with oil representing one-fifth of the total by 1925 and rising to one-third by World War II.[37] By 1940, the United States accounted for over two-thirds of world production. Until mid-century, the United States exported oil to its friends and allies. In classical economic theory, demand drives supply; in the case of oil, however, controlling the supply supported prices at levels that made it profitable to invest in drilling new wells. Unchecked supplies of domestic oil caused collapsing prices. To counter this trend, modern mobility was shaped by the domestic abundance of oil. This included where we lived in relationship to where we worked, the clothes we wore and where they were made, what and where we ate and our need to travel.

In January 1948, James Forrestal, the U.S. secretary of defense, warned that unless United States gained and maintained access to Middle Eastern oil, consumers would have to accept four-cylinder engines for their automobiles rather than the V-8 engines then in production. In that year, per capita consumption of oil was 14.4 barrels. In 2010, it had skyrocketed to 22.6 barrels.

> Had U.S. public policy through the preservation of public transportation, the promotion of efficiency, and other measures (including four-cylinder engines), maintained the 1948 level of oil use, U.S. oil consumption in 2010 would have been almost 40 percent lower, with consequent benefits for the economy, U.S. national security, and the environment.[38]

Consumption became a national trend. U.S. automobile manufacturers abandoned more fuel-efficient six-cylinder internal combustion engines at a time in which European automakers were developing successful four- and two-cylinder vehicles. The new V-8 engine, popularized in the United States in the 1950s, placed consumers on a heavy-carbon path of consumption and ensured big oil firms a stable market for gasoline. Scientific breakthroughs in the refining process allowed crude to be separated into different products. Gasoline refined from a barrel of crude grew from 15 percent in 1900 to 39 percent in 1929 and continued to rise in the following decades.[39]

Iraq

In the Middle East, the Great War ended with the destruction of the Ottoman Empire, Germany's overseas empire and Austria-Hungary. The division of its provinces and regions into nation-states adhered to the decisions made at the Treaty of Paris following the war. Its Middle East territories were passed on to British and French forces that controlled them for the next five decades. During the war, British forces invaded the territory near Basra and by 1918 captured Baghdad. They controlled Mesopotamia's (Iraq's) oil riches. Anglo-Iranian Oil Company would become British Petroleum (BP). It negotiated with a consortium, the Iraqi Petroleum Company (IPC) that included Shell Oil and U.S. firms and the French Compagnie française des pétroles (today known as Total). IPC would remain in command of the region's oil for more than 50 years. In January 1968, the British government withdrew its decades-long political and military power from the Middle East. The void created by Britain's decision would be filled by the U.S. presence in the region.

Decades before the pullout by the British, the then newly created Iraqi government under its king, Emir Faisal in 1933, needing a share of the royalties from oil exploration ceded control of its primary economic resource to the consortium that included Anglo-Iranian and U.S. oil firms, including the most powerful, Standard Oil. By acquiescing to this arrangement, the new Iraqi kingdom gained control of the oil-rich province of Mosul rather than see it ceded to the government in Turkey. Drilling began in April 1927, and within weeks, workers discovered a vast reservoir of oil north of the city of Kirkuk. The IPC limited production to 2,000 barrels a day. The onset of the Great Depression in 1929 curtailed production further. Needing to increase its revenue through added oil production, the Iraqi government agreed to eliminate the tax on the company's profits and expand the company's drilling area from 192 square miles to 32,000 square miles.

With further concessions a decade later, BP and Total built two 12-inch pipelines from Kirkuk to the Mediterranean ports at Haifa in Palestine and Tripoli in Syria. At the time, it was the biggest welded pipeline in the world and increased oil production from 2,000 to 80,000 barrels a day. Although Iraq maintained a small refinery for processing oil for domestic consumption, it remained dependent on the consortium controlled by BP for processing oil for export. By 1950, production doubled again to 160,000 barrels per day and doubled again in 1952. By 1980, production reached 2.5 million barrels per day.[40]

In the interim, during the 1950s and the 1960s, increases in production remained at about half that of Iran, Saudi Arabia and Kuwait. BP and its partners limited production to avoid a glut on the export market that would depress prices. It engaged in now-familiar strategies, "deliberately drilling shallow wells to avoid discovering additional supplies, and pulled wildcat wells that yielded large finds to conceal their existence from the government."[41] The overthrow of the British-supported monarchy by army officers led by Abd al-Karim Qasim in 1958 was inspired, in part, by the desire to nationalize the country's oil reserves.

With Iraqi oil controlled by the political, military and technical might of foreign companies and their governments, the new nationalist government's efforts to convince IPC to construct a pipeline from the Mosul oilfields to refineries in Basra in the south for export failed. In further negotiations, the nationalist government was prepared to accept IPC ownership of existing wells but waved the right to negotiate with other oil companies for drilling rights in areas with unproven reserves. The British Foreign Office feared that Iraq might then annex Kuwait, a former dependency of Basra Province. Fast-forward to the first Iraqi War (1990–1991) in which Saddam Hussein invaded Kuwait claiming it as a province of Iraq.

Failed negotiations led in December 1960 to the cancellation of the agreements that conceded known reserves to IPC but then denying access to reserves that they refused to develop. BP shut down Iraqi oil production and eliminated the threat of overproduction from the region's drilling and refinery operations. A new military government established a state-owned company, Iraqi National Oil in 1964 with financial support from the Soviet Union. It built a state-owned pipeline to Basra to refine oil from its known reserves and the large new oil fields in North Rumaila. In July 1968, military officers affiliated with the Ba'ath Party and led by Saddam Hussein engineered another coup and built the refineries needed to process and export its state-controlled oil. On June 1, 1972, the Ba'athist government nationalized IPC. It was the first Middle Eastern country to eliminate control of its oil reserves from British and American domination.[42]

Saudi Arabia

The oil fields of Saudi Arabia became the domain of four U.S. companies that owned the Trans-Arabian Pipeline Company. In 1933, Standard Oil of California (Chevron) joined by Texas Oil (Texaco) in 1936 to create Aramco, a joint venture. Its pipeline passed through Syria to a terminal on the coast of Lebanon, near Sidon. Like the concessions granted to IPC, Aramco operated free of royalty payments for oil, taxes, and import duties. It laid pipelines without restrictions and used natural resources, wood, stone and water, with impunity. The discovery of massive oil fields in Saudi Arabia by Aramco in 1938, "the greatest geological prize the world has ever known,"[43] shifted the world's oil production from the Americas to the Middle East. To raise enough financial capital and to spread the risks of developing the massive reserves, Chevron and Texaco invited Standard Oil of New Jersey (Exxon) and Socony (Mobil) in 1947 to join Aramco.

By the 1940s, oil's increasing importance to fuel the war machines of the Allies and Axis Powers alarmed both sides about its continuing availability during World War II and in the postwar period. The failure of the Axis Powers (Germany, Japan and Italy) to supply the increasing demands for oil played a significant role in their defeat. To acknowledge the significance of the shift to the Middle East in oil production, President Roosevelt met with the Saudi king, Adb al-Aziz Saud, the founding king of the country on the U.S.S. *Quincy* on Egypt's Great Bitter Lake in February 1945 before returning home from the four-power meeting at Yalta in southern USSR.

An increasingly mobile consumer society strained domestic supplies in the postwar period. The reconstruction of Europe in the aftermath of the war required foreign sources of oil. The government of Franklin Delano Roosevelt (1933–1945) recognized that the Middle East represented the new "center of gravity" for world oil production. In 1948, Aramco began drilling into the Ghawar field in eastern Saudi Arabia. The oil field turned out to be as large as the state of Delaware with reserves of at least 80 to 100 billion barrels of crude. It was 10 times the size of the immense East Texas oil field, the largest in the United States. Cheap Saudi oil provided Europe, dependent on Marshall Plan, assistance in recovering from the ravages of World War II with an energy supply to rebuild its homes, factories and infrastructure. It also triggered the beginning of Europe's long-term dependence on Middle Eastern oil.[44]

Iran

Protecting the security and stability of the Middle East to ensure an uninterrupted flow of oil to oil-dependent Europe and heightened consumer demand in the United States became foreign policy imperatives. Saudi Arabia, Iraq and Iran became focal points in U.S./Middle East policy. British, Soviet and U.S. military forces occupied Iran during World War II to prevent Nazi Germany from gaining access to its oil. The end of the war led to the withdrawal of Allied forces and the creation of a constitutional monarchy, with Mohammed Reza Shah Pahlavi, the shah of Iran, as monarch. In 1945, the Middle East produced only 7.5 percent of the world's oil, with Anglo-Iranian oil producing two-thirds of that amount.[45]

Led by the country's Nationalist prime minister, Mohammed Mossadegh, the parliament voted to nationalize the British-owned Anglo-Iranian Oil Company in 1951. As Britain's key asset in the region, fear that Iran's decision might create a wave to nationalize oil production throughout the region and jeopardize the 50–50 profit-sharing arrangement between oil firms and Middle Eastern countries. Efforts by the United States to negotiate an agreement, short of nationalization failed, as did efforts to convince the shah to remove the prime minister.

Fear that the Iranian elected government under the leadership Mohammad Mossadegh favored an alliance with the Soviet Union resulted in a coup in 1953 organized by the Central Intelligence Agency and British intelligence deposing Mossadegh. The shah of Iran established a royal dictatorship, destroyed parliamentary governance and promised to protect foreign oil interests and suppress support for unionization. Demands for better working conditions, higher wages and the right to unionize were met with force.[46]

With growing dependence in Europe, Japan and the United States on oil from the Middle East, preserving the security of the autocrats in power, protecting the international cartel of oil companies drilling and transporting to its clients became paramount considerations. Maintaining price stability (not a low) and not allowing the world's second-most abundant fluid to reach the market unchecked would prevent prices from collapsing. Maintaining scarcity would keep client Middle Eastern countries stable and guarantee large profits for the oil companies.[47]

Price stability and protecting governments aligned with U.S. strategic interests in the region supported massive arms sales to Iraq, Iran, and Saudi Arabia. Autocratic governments received money from oil contracts and used a measurable sum to purchase advanced weapons systems, including small arms, ground vehicles, aircraft and computers used for reconnaissance and surveillance. From 1970 to 1979, the United States sold US$22 billion worth of arms to Iran and, beginning in 1972, US$3.5 billion to Saudi Arabia, with a much smaller population.[48] The

Soviet Union sold US$10 billion in arms to its client state, Iraq. The militarization of the region, intended to create stability, proved to be an illusion.

Arab–Israeli conflict

When the conflict began, its origins may be traced to events millennia ago. However, the decision to partition Palestine and establish the state of Israel in 1947 by the United Nations' Special Committee on Palestine angered the Arab states and Iran. The withdrawal of British troops in 1948 set the stage for a bitter conflict among Israeli settlers and the Arab League. On May 14, 1948, the Jewish National Council officially pronounced the establishment of the state. Recognition by the Soviet Union and the United States was followed by an attack by the Arab League (Lebanon, Iraq, Jordan and Egypt) against Israel. Both sides signed an armistice in 1949, with the West Bank and Gaza held by Egypt and Jordan and with Israel gaining additional Palestinian lands. The truce held until 1967 when the Palestinian organization Fatah and other Palestinian guerilla organizations invaded Israel. With Egypt, Jordan and Syria providing military support for the guerillas and entering the fight against Israel, in the five days, from June 5 to 10, Israel defeated the opposition and extended its occupation to include the Sinai Peninsula, the Gaza Strip, East Jerusalem and the Golan Heights.

United States support for Israel required a delicate balancing of its commitment to Israel and its support for the oil producers in Arab states and Iran, at a time when the U.S. allies in Europe and Japan depended on oil from the region. Soviet support for Egypt and Iraq through arms sales required an Israeli-friendly foreign policy from the United States. Egypt's surprise attack on Israel in October 1973 to recover territory in the Sinai Peninsula upset the unstable balance. On appeal, the United States helped to reequip Israel's military, a decision that infuriated the Arab states.

Petro-dollars and weapons purchases

In retaliation to U.S. arms support to Israel, the embargo on oil in 1973 destined for the United States meant that seizing control of production and pricing would increase their profits. Led by Saudi Arabia with support from Kuwait, Iran and the other Organization of Petroleum Exporting Countries (OPEC) countries, corporate control by major oil companies of the reserves in Middle Eastern countries ended with the nationalization of those reserves. OPEC formed in 1960 pressured Aramco and others to increase its profit-sharing with them. The embargo took the long-term corporate strategy of controlling prices by withholding production to its extreme. As a result, shortages and long lines at service stations in the United States traumatized American consumers. The spike in gasoline and heating oil prices created an inflationary spiral in the United States and newfound wealth for OPEC. Much of it found its way back to the United States in increased weapons sales to support autocrats in the region.

Although it intended to bolster regional security, its unintended effect emboldened regional dictators to assert their power and "rattle their weapons." This outcome was especially true in the historic conflict between Shia and Sunni Muslims and, particularly, between Iraq and Iran. "Between 1975 and 1979 Iran, Iraq and Saudi Arabia purchased 56 percent of all weapons sold in the Middle East and almost one-quarter of all global arms sales."[49] The new oil wealth did not result in the democratization of the political systems. Social service reforms and subsidies followed to satisfy growing populations in the Middle East. Universal health care and compulsory schooling created a more literate and physically healthy population. Housing subsidies ameliorated substandard living arrangements. In fact, higher literacy rates and improved well-being may have enlightened people about opportunities in other parts of the world but frustrated them regarding the absence of political freedoms. In all cases, brutal tactics imposed by dictators in the region suppressed political expression. According to an Amnesty International assessment of the Shah of Iran's governance in 1977, it has "the highest rate of death penalties in the world, no valid system of civilian courts and a history of torture which is beyond belief. No country in the world has a worse record in human rights than Iran."[50]

Oil sales and the purchase of military weapons became intertwined. As Middle Eastern regimes either nationalized production or demanded a larger share of profits from the Anglo-American cartel, pounds sterling and dollars flowed into the treasuries of OPEC. Purchasing food and consumer durables, such as cars, household equipment, machinery and commodities, for infrastructure modernization, real estate purchases in the United States and the United Kingdom could not begin to deplete the hoards of foreign currency held in Middle Eastern treasuries. The surplus of petro-dollars from Saudi Arabia was used to buy U.S. Treasury securities as another way to recycle dollars back to the United States and continue the practice of pricing oil exclusively in dollars.

> To maintain the balance of payments and the viability of the international financial system, Britain and the United States needed a mechanism for these currency flows to be returned . . . Arms were particularly suited to this task of financial recycling, for their acquisition was not limited by their usefulness.[51]

About US$500 billion was recycled back to the U.S. using the methods described above.

Since autocratic governments, especially Saudi Arabia, Iran and Iraq, could justify their purchases on the grounds that the region was inherently unstable and more purchases could be justified on the grounds of national security, the American government and arms manufacturers became willing participants in the pattern of recycling petro-dollars. As weapon systems became more technically complex and attack fighter aircraft became more expensive, costing tens of millions of dollars per aircraft, Saudi Arabia, Iraq and Iran purchased such aircraft in increasing numbers.

The Iranian Revolution of 1979

The revolution that deposed the shah and replaced his autocratic regime with an autocratic fundamentalist theocracy led by the cleric Ayatollah Khomeini disrupted U.S. policy in the region turning an ally into a menacing adversary. The assault and capture of American embassy personnel in the capital city, Tehran lasted 444 days in 1980–1981. Not until 2015 were both countries prepared to sign an agreement to limit Iran's nuclear capability in return for the United States lifting of economic sanctions. Sensing a menacing theocratic Shia rival in Iran, Iraq's Ba'athist and Sunni dictator, Saddam Hussein, attacked Iran's oilfields in September 1980. The war lasted until 1988, killing and wounding hundreds of thousands, probably 1 million combatants and civilians. Hussein's motives are unknown, but a patriotic war would cover his domestic failures and perceived threats from others in the region. Large stockpiles of weapons purchased with petro-dollars provided him with the wherewithal to wage a protracted war. The United States supplied both sides in the conflict with intelligence and more weapons in the hopes that it would prevent the two most militarized countries in the region from gaining primacy and disrupting the flow of oil from other OPEC nations in the Middle East. The United States also knew that Hussein used chemical agents on the battlefield against Iran.[52]

The decision to finance and equip the combatants by the United States involved it militarily in Middle Eastern politics, an involvement that would continue into the twenty-first century. During the Iraq–Iran War, one OPEC member, Kuwait, in 1986, appealed to the United States and the Soviet Union to protect its oil tankers from attacks by Iran. The United States responded by permitting the Kuwaiti tankers to fly the American flag, signaling to Iran that an attack on these ships was an attack on the United States. Armed conflict between the United States and Iran escalated soon thereafter. The United States sent a naval fleet to the region to protect its oil interests, where it remains today. Iranian and U.S. gunships exchanged fire several times with the U.S. ships sinking several Iranian vessels and damaging many of its oil platforms in the Persian Gulf. On July 3, 1988, the U.S. cruiser, the U.S.S. *Vincennes*, shot down an Iranian Airbus A300 passenger jet, killing all 290, including 66 children. The enmity directed toward the United States by Iran and the U.S. support for Iraq during the war remain to this day. Weakened by the war, Iran accepted a United Nation's ceasefire that same year.[53]

Although the war ended, its aftermath revealed a weakened Iraq, deep in debt to other OPEC nations who helped to finance this protracted debacle. Creditors in the Arab world demanded repayment as they flooded the oil market with their excess. Rebuilding Iraq's oil infrastructure and returning it to prosperity required stable (not low) oil prices. Unable to meet demands at home for a return to domestic peace and repayment of his war debts and a military weakened by eight years of war but not defeated, Saddam Hussein exercised the military option once again. In August 1990, he invaded Kuwait, which had been a protectorate of

Britain until 1961 and an independent nation thereafter. It gained admission to the United Nations in 1963. It also continued to pump cheap oil onto the international market, causing a further erosion of prices.

Operation Desert Storm (1990–1991) and the Iraq War (2003–2011)

Already involved militarily in Middle Eastern economics and politics, the United States became more deeply entrenched in the region by leading an international coalition and sending 500,000 troops into the breach. Within days, the coalition expelled Iraq's invading troops and destroyed its retreating army. A defeated Iraq with economic sanctions imposed by the United States and approved by its coalition partners gutted the country's economy and further weakened its social fabric.

After the September 11, 2001, attack on the World Trade Center in New York City, led by jihadists from Saudi Arabia, the United States used the pretext of Iraq's possession of nuclear materials and weapons to invade the country with few coalition partners. The defeat of the Ba'athist government, the demobilization of the Iraqi armed forces and the execution of Saddam Hussein by the new government did little to create a peaceful environment. Sectarian violence between Sunni and Shi'ite militias and Kurdish separatists destabilized the country. The balance of power that existed in the region prior to the invasion disappeared, with Iran becoming a more aggressive sponsor of terrorist activity in the Middle East.

> If oil and American oil policy are kept in focus, this period constitutes not a series of wars, but a single long war, one in which pursuing regional security and protecting oil and American-friendly oil producers has been the principal strategic rationale.[54]

The militarization of the Middle East and the single long war to keep oil flowing for the benefit of the United States and its allies has come about at great cost in terms of human suffering and casualties. The massive U.S. military presence in the Persian Gulf from 1976 to 2007 was estimated to be US$7 trillion while the cost of the Iraq War (2003–2011) added another US$3 trillion.[55] Part of the reason for these enormous costs was the regional instability caused by revolutions and wars. The Iranian revolution that began in 1979 and the embargo that followed caused the second sticker price shock in the 1970s. Motorists in the United States aided the shock by "hoarding fuel, idling engines in gas lines, and frantically topping off their tanks with frequent trips to the local filling station."[56]

With sanctions placed on Iran in the wake of its theocratic revolution, expectations that high consumption (per capita consumption in the United States rose from 243 gallons in 1950 to 463 gallons in 1979) would result in higher prices at the pump. However, oil exploration in the deep water Gulf of Mexico and Alaska's Prudhoe Bay and the decision by Saudi Arabia in 1985 to abandon OPEC pricing and flood the market with cheap oil to gain market share caused a collapse in prices. For the next 14 years, prices fluctuated at low levels, declining to US$14 a barrel in 1986. Manufacturers increasing the supply of sport utility vehicles and small trucks for sale satisfied America's thirst for large, fuel-inefficient motor vehicles. Arguably, the collapse of the Soviet Union's economy dependent on its international sale of crude oil led to the disintegration of the Union between 1989 and 1991. Losses there accumulated to the tune of US$20 billion a year.[57]

With the Soviet Union's demise and the end of the Cold War, U.S. military presence in the Middle East continued to protect its access to oil. Cheap oil prices did not continue, however, as state-owned national oil companies (NOCS) and OPEC nations cut production to increase prices. Petroleos de Venezuela, Gasprom in Russia and Petrobras in Brazil as NOCS and OPEC gained a larger share, upward of 80 percent of all known oil reserves and an increasing share of the world's refining, transport and marketing of oil. From a low of US$14 a barrel in 1986, cutting production over the long-term saw prices reach a high of US$147 per barrel in 2008.[58] Having lost control over the global supply of oil, the United States relaxed antitrust enforcement and allowed the major oil firms to merge, with BP, ExxonMobil, Royal Dutch Shell, Chevron and ConocoPhillips leading the way. In addition, the government relaxed regulatory oversight and permitted increased drilling offshore in the deepest waters of the Gulf of Mexico. The blowout of the Macondo oil well and the death and injury of drillers on BP's *Deepwater Horizon* vessel in 2010 reflected the consequences of weak oversight.

Offshore drilling did not reverse the downward trend in U.S. domestic production in the new century. By 2005, the country imported 60 percent of its oil as consumption exceeded the reduced supplies made available by NOCS and

Figure 4.2 Modern offshore oil drilling platform with supporting supply ship.

OPEC nations. Having lost control of pricing, for a nation that seemed "addicted to foreign oil," the future looked rather bleak. At the same time, energy analysts predicted a decline in natural gas reserves. In response, energy producers began building liquefied natural gas facilities to compensate for eventual shortages.

Conclusion

Predictions about future oil and gas shortages in the United States proved to be erroneous due to the failure to account for advances in production technology. Hydraulic fracturing of oil-saturated shale located miles below Earth's surface created an oil industry boom. Domestic oil production has skyrocketed in the last six years by more than 70 percent, to 12 million barrels a day, eclipsing Saudi Arabia's output of 11.5 million barrels. Some grades of oil, accounting for about one-third of all the oil Americans use will continue to be imported by the United States for years to come because its refining capacity was built to process heavier crude from Mexico, Venezuela and Canada. By contrast, shale oil is light, with a lower sulfur content that allows it to flow at room temperature. Unheard of just years before, in 2011 the United States became one of the world's leading exporters rather than one of the world largest importers of oil, a condition not experienced in more than a half century.

Exporting oil changes U.S. relationships with the rest of the world, especially the Middle East. As energy historian Daniel Yergin has pointed out,

> [e]conomically, it means that money that was flowing out of the United States into sovereign wealth funds and treasuries around the world will now stay in the U.S. creating jobs . . . it certainly provides a new dimension to U.S. influence in the world.[59]

This monumental shift in oil production from foreign to domestic sources altered the geopolitical landscape. As global prices for crude began their steady decline, many OPEC nations, especially Saudi Arabia and Kuwait, continued to pump oil to maintain market share and drive more expensive producers in the United States from the global market. As prices dropped to US$30 a barrel in 2016, this tactic only accelerated the steep decline in prices without the intended benefit. For American consumers who consumed 1,200 gallons of gasoline on average each year, the decline represented huge savings in fuel costs. With 94 million barrels a day entering the global market, oil-producing countries that depend entirely on this revenue faced huge financial deficits.

Since OPEC nations benefit primarily from stable (not low) oil prices, its power to negotiate and set international prices collapsed. In 1973, OPEC triggered the oil embargo in response to U.S. military support for Israel, attacked by Egypt to regain territory lost to Israel during the 1967 war. Crude rose from US$3 a barrel to US$12 and resulted in OPEC wresting control over global oil production and prices from the international oil companies. Higher oil prices became the norm as OPEC ministers met semiannually in Vienna, Austria to either increase or withhold production to stabilize high prices. Geopolitical circumstances have now changed with collapsing oil prices. Most OPEC nations need between US$100 and US$80 per barrel of crude to maintain their budgetary needs. Iran, which recently emerged from a long era of sanctions imposed by the United States and its allies, was forced to cut production from 1979 to 2016. Once it signed an agreement with the United States and others to stop producing weapons-grade plutonium, it was permitted to raise oil production by about 1 million barrels a day. Iran needs oil priced at US$147 a barrel to maintain its budgetary needs. With the United States now becoming a net exporter of oil, it was able to convince China and India to reduce its oil imports from Iran. This "hardball" strategy may have played a role in bringing Iran to the table to negotiate a freeze of its nuclear program.[60] Relations between Iran and the United States have deteriorated now that President Donald Trump has withdrawn from the nuclear arms agreement signed by the administration of Barack Obama (2009–2017).

Notes

1 Michael B. McElroy, *Energy: Perspectives, Problems, & Prospects* (New York: Oxford University Press, 2010), 128.
2 M. Polo, *The Book of Ser Marco Polo: The Venetian Concerning the Kingdoms and Marvels of the East* (London: John Murray, 1903), 46.
3 Oliver Kuhn, "Odyssey of Oil," *Recorder: The Canadian Society of Exploration Geophysicists*, 25, no. 10 (December 2000): 5–7.
4 Zhong Changyong and Huang Jian, *Drilling and Gas Recovery Technology in Ancient China* (Wictle Offset Printing, 1997).
5 Paul Lucier, *Scientists & Swindlers: Consulting on Coal and Oil, 1820–1890* (Baltimore, MD: The Johns Hopkins University Press, 2008), 191.
6 Ibid., 200.
7 Ibid., 202.
8 Brian Black, *Petrolia: The Landscape of America's First Oil Boom* (Baltimore, MD: The Johns Hopkins University Press, 2000), 32.
9 Ibid., 34–35.
10 Ibid., 60. From *The New York Times*, December 20, 1864, 5.
11 Christopher F. Jones, *Routes of Power: Energy and Modern America* (Cambridge, MA: Harvard University Press, 2014), 96.
12 Ibid.,100.
13 Timothy Mitchell, *Carbon Democracy: Political Power in the Age of Oil* (New York: Verso, 2011), 36.
14 Ibid., 36–37.
15 Jones, *Routes of Power*, 159.
16 Ibid., 108.
17 Black, *Petrolia*, 66. As quoted in *Harper's* 50 (1865): 54.
18 Lucier, *Scientists & Swindlers*, 211.
19 https://upload.wikimedia.org/wikipedia/commons/6/6e/Crude_Oil_Distillation-en.svg.
20 Lucier, *Scientists & Swindlers*, 232–233.
21 Black, *Petrolia*, 138.
22 Daniel Yergin, *The Prize: The Epic Quest for Oil, Money, and Power* (New York: Simon & Schuster, 1991), 39.
23 Jones, *Routes of Power*, 106.

24 Ibid., 112.
25 Yergin, *The Prize*, 56.
26 Ibid., 61–62.
27 Mitchell, *Carbon Democracy*, 32–33.
28 Ibid., 46.
29 Ibid., 46–47.
30 Paul Sabin, *Crude Politics: The California Oil Market, 1900–1940* (Berkeley: University of California Press, 2005), 3–6.
31 Yergin, *The Prize*, 84.
32 Ibid., 85.
33 Mitchell, *Carbon Democracy*, 46–47.
34 Ibid., 64.
35 David S. Painter, "Oil and the American Century," *Journal of American History*, 99 (June 2012), 24.
36 Brian Black, *Crude Reality: Petroleum in World History* (Lanham, MD: Rowman & Littlefield, 2012), 79–80.
37 Painter, "Oil and the American Century," 25.
38 Ibid., 38.
39 Black, *Crude Reality: Petroleum in World History*, 81.
40 Ibid., 102–103.
41 Mitchell, *Carbon Democracy*, 147.
42 Ibid., 151.
43 Tyler Priest, "The Dilemmas of Oil Empire," *Journal of American History* 99 (June 2012): 237.
44 Ibid., 238.
45 Mitchell, *Carbon Democracy*, 113–114.
46 Painter, "Oil and the American Century," 24–31.
47 Toby Craig Jones, "America, Oil, and War in the Middle East," *Journal of American History* 99 (June 2012): 208. Mitchell, *Carbon Democracy*.
48 Jones, "America, Oil, and War in the Middle East," 212.
49 Ibid., 213.
50 Martin Ennals as quoted in Noam Chomsky and Edward S. Herman, *The Washington Connection and Third World Fascism: The Political Economy of Human Rights*, vol. 1 (Cambridge, MA, 1979), 13. Also quoted in Jones, "America, Oil, and War in the Middle East," 214.
51 Mitchell, *Carbon Democracy*, 155.
52 Jones, "America, Oil, and War in the Middle East," 215 citing the work of Joost R. Hiltermann, *A Poisonous Affair: America, Iraq, and the Gassing of Halabja* (New York: Cambridge University Press, 2007), 37–64.
53 Jones, "America, Oil, and War in the Middle East," 215–216.
54 Ibid., 216.
55 Roger J. Stern, "United States Cost of Military Force Projection in the Persian Gulf, 1976–2007," *Energy Policy* 38 (June 2010): 2816–2825. Joseph E. Stiglitz and Linda Bilmes, *The Three Trillion Dollar War: The True Cost of the Iraq Conflict* (New York: W.W. Norton, 2008).
56 Priest, "The Dilemmas of Oil Empire," 242.
57 Ibid., 245.
58 Ibid., 246.
59 Clifford Krauss, "Reversing the Flow of Oil," *The New York Times*, Special Section "Energy" (October 8, 2014), F1, F6.
60 David Wallis, "Oil's Comeback Gives U.S. Global Leverage," *The New York Times* Special Section "Energy" (October 8, 2014), F7.

References

Black, Brian. *Crude Reality: Petroleum in World History*. Lanham, MD: Rowman & Littlefield, 2012.

Black, Brian. *Petrolia: The Landscape of America's First Oil Boom*. Baltimore, MD: The Johns Hopkins University Press, 2000.

Changyong, Zhong, and Huang Jian. *Drilling and Gas Recovery Technology in Ancient China*. Hong Kong: Wictle Offset Printing, 1997.

Hiltermann, Joost R. A *Poisonous Affair: America, Iraq, and the Gassing of Halabja*. New York: Cambridge University Press, 2007.

Jones, Christopher F. *Routes of Power: Energy and Modern America*. Cambridge, MA: Harvard University Press, 2014.

Jones, Toby Craig. "America, Oil, and War in the Middle East." *Journal of American History* 99, no. 1 (June 2012): 208–218. doi: 10.1093/jahist/jas073.

Krauss, Clifford. "Reversing the Flow of Oil." *The New York Times*, October 8, 2014, F1, F6.

Kuhn, Oliver. "Odyssey of Oil." *Recorder: The Canadian Society of Exploration Geophysicists* 25, no. 10 (December 2000): 5–7.

Lucier, Paul. *Scientists & Swindlers: Consulting on Coal and Oil, 1820–1890*. Baltimore, MD: The Johns Hopkins University Press, 2008.

McElroy, Michael B. *Energy: Perspectives, Problems, & Prospects*. New York: Oxford University Press, 2010.

Mitchell, Timothy. *Carbon Democracy: Political Power in the Age of Oil*. New York: Verso, 2011.

Painter, David S. "Oil and the American Century" *Journal of American History* 99, no. 1 (June 2012): 24–39. doi: 10.1093/jahist/jas073.

Polo, M. *The Book of Ser Marco Polo: The Venetian Concerning the Kingdoms and Marvels of the East*. London: John Murray, 1903.

Priest, Tyler. "The Dilemmas of Oil Empire." *Journal of American History* 99, no. 1 (June 2012): 236–251. doi: 10.1093/jahist/jas073.

Sabin, Paul. *Crude Politics: The California Oil Market, 1900–1940*. Berkeley: University of California Press, 2005.

Stern, Roger J. "United States Cost of Military Force Projection in the Persian Gulf, 1976–2007." *Energy Policy* 38, no. 6 (June 2010): 2816–2825. doi: 10.1016/j.enpol.2010.01.013.

Stiglitz, Joseph E., and Linda Bilmes. *The Three Trillion Dollar War: The True Cost of the Iraq Conflict*. New York: W.W. Norton, 2008.

Wallis, David. "Oil's Comeback Gives U.S. Global Leverage." *The New York Times*, October 8, 2014, F7.

Yergin, Daniel. *The Prize: The Epic Quest for Oil, Money, and Power*. New York: Simon & Schuster, 1991.

5 A history of manufactured gas and natural gas

Introduction

Manufactured gas became one of the first networked production and distribution enterprises along with water and sewer systems of nineteenth-century cities. It preceded by decades the development of technological systems such as trolley, telephone and electric systems to create an integrated infrastructure. Paving city streets had a similar networking effect. Manufactured gas lasted for 170 years before it was replaced by natural gas. Its history began with mining coal, transporting it by wagon, barge and eventually railway. It required combustion without oxygen to produce coke, in a retort (a sealed oven). The retort captured the coal gas released from the production of coke. Purification and storage came last in this process but before transportation. Transporting the purified gas through a network of pipes to streets, commercial buildings and the homes of elites for lighting transformed urban life in ways unimaginable in the decades before.

With workers no longer restricted to the limitations imposed by oil lamps, candlelight or daylight hours, gas lighting extended the workday. It illuminated previously darkened streets, encouraged sociability and promoted safety. Oil lighting of the streets of London at night was "not only dismal but hardly enabled the passenger to distinguish the watchman from the thief or the pavement from the gutter."[1] Gas lighting changed people's perceptions of the night. "The light is but little inferior to daylight and the streets are divested of many terrors and disagreeables."[2]

The process of carbonization

Burning coal produced more than 80 percent of the town gas manufactured in Britain and the United States, countries with large coal reserves. At the height of distillation, producers heated as much as 34 million tons of coal and 1.5 tons of coke annually. To carbonize coal, mechanical crushers reduced its size to 1-inch diameters or less and shoveled it into a sealed oven, a retort made of firebricks or silica. The retort, once heated to temperatures ranging from 600 degrees Celsius, or 1112 degrees Fahrenheit, softened the coal. Each ton of coal heated in this way yielded 14,000 cubic feet, or 550 British Thermal Units (BTUs), of gas; 1,120 pounds of coke; 10 gallons each of tar and sulfur; and 28 pounds of ammonia.

As historian Leslie Tomory has noted, eighteenth-century Enlightenment scientists without a particular application in mind explored the characteristics of pneumatic chemistry, the chemistry of air that led to the discovery of oxygen, hydrogen, carbon monoxide, carbon dioxide and many others. They knew the chemical properties of lime as a purifying agent, and their laboratories housed retorts, condensers and gasometers (storage tanks), all necessary equipment in the production of manufactured gas. Producers used wood chips and iron filings to eliminate the bad odors caused by using lime. Controlled scientific experiments resulted in findings that suggested the best coals to use for carbonization, effective temperatures and the best sizes and shapes of retorts. With this creative practical application of scientific knowledge, the Industrial Revolution flourished. As a result, the Scientific Revolution and the Industrial Revolution became entwined.[3]

The growth of industrial distillation in the late 1700s reflected not only the combination of scientific and technological knowledge but also dire environmental conditions facing the economies of Britain and those in France and Germany. In the case of manufactured gas, industrial distillation is a separation technology designed to make a more efficient use of energy sources. Tightly controlled distillation ovens produced coked coal and charcoal by burning coal and wood. Burning coal also produced tar and pitch, two products needed as sealants for ships. Canals transported coal to their industrial customers before the development of railroads. The loss of its colonies in the Americas caused shortages in the British navy affecting its global empire. The depletion of forests in continental Europe, beginning in the late Middle Ages accelerated in the centuries that followed. Turning wood to charcoal and capturing its gas coal became a way to foster a more efficient energy regime.

"The crude gas consists of yellowish-brown smoke up to (176° F.) 80° C. A large purification plant accepted the condensed crude gas and passed it through a set of tubes cooled by water and air to (59° F.) 15° C. removing tar and water vapors. Washing and scrubbing the pure gas continues as water and chemicals including lime were used in a large vessel containing grids across a large wetted surface moving against the flow of the gas. Once completed, gasholder facilities stored the finished product. Such gasworks contained pumps called exhausters to facilitate the flow of gas to customers and meters to record the volume sent to municipalities for gaslights, to factories and households."[4]

Coal gas is produced from the carbonization of coal and consists of hydrogen, methane, carbon monoxide, and illuminants. Gasification processes and the associated fuel gas produced included: producer gas consisting of carbon monoxide and hydrogen; carburetted water gas consisting of hydrogen, carbon monoxide and illuminants, and oil gas consisting of hydrogen and methane with a lesser amount of carbon monoxide and illuminants. Each of the town gas processes discussed (i.e., coal gas, oil gas and carburetted water gas) contained water vapor, tar, naphthalene and hydrogen sulfide as impurities.

The removal of these impurities resulted in residuals containing the following components. Water vapor was removed as condensate. Tar-water emulsions were common to oil gas and carburetted water gas processes that used heavy oils or asphaltic-based oils. Emulsions were not common from processes that used lighter oils or from coal gas processes. Much of the naphthalene would be removed with the tar. In the oil gas process, naphthalene was removed using a wash oil and the used wash oil was recycled into the gas-making oil. Town gas purifier wastes generally contained sulfides, but cyanide-bearing purifier wastes were common only to coal gas processes. Lampblack was generated in significant quantities only from the oil gas process.[5] All of these complicated processes released contaminates that fouled the land and polluted the atmosphere.

British and U.S. origins

Labeled a "smokeless fuel," its boosters, particularly in Britain where households, factories and ships burned millions of tons of coal for fuel promoted as a way to make the city's country air cleaner. The first public gasworks began operating in Britain in 1813 and continued to produce manufactured (town) gas until the discovery of natural gas in the North Sea in the 1960s. Located in Great Peter Street in Westminster and owned by Gas Light and Coke Company, its success in delivering town gas to its customers encouraged others to enter the business of production. By 1814, 122 miles of pipelines supplied 248 million cubic feet of town gas to 31,000 street lamps by burning 50,000 tons of coal in retorts.[6] By 1820, 15 major towns in England and Scotland possessed gasworks. Ten years later, 200 towns dotted Great Britain. In less than 50 years, 760 towns in Britain and Ireland owned at least one gasworks. Since delivery required a network of pipes and mains, citizens experienced a degree of disruption as workers dug up city and town streets to lay pipes and mains to connect gas to street lights, public and commercial buildings, factories and residences. "Between 1812 and 1820, London's gas infrastructure grew from nothing to a point where gas mains reached most areas of the city."[7]

By 1850, London contained 2,000 miles of gas lines. With Britain's enormous coal reserves, much was devoted to producing manufactured gas, a relatively new industry. The rise in coal use reflected the considerable growth in the manufactured gas industry from 500,000 tons in 1830 to 10 million tons in 1887. London consumed the largest share of the industry's coal, amounting to 4 million tons each year. Throughout the nineteenth and the early years of the twentieth centuries, town gas illuminated most of Britain.

Individual acts of Parliament gave companies proprietary rights to build gasworks on specific lands, to dig up streets and to lay pipelines. In return, companies pledged to guarantee high-quality gas to customers at fixed prices and to pay a dividend to its stockholders. Future Gasworks Clauses Acts updated these conditions and included new provisions for official testing for quality, purity and pressure. By 1882, the industry maintained its 70-year monopoly over gas lighting with an annual output of 65 billion cubic feet. Despite the monopoly, thousands of independent gas producers located in many villages and towns not linked into the main gas system locations delivered manufactured gas to customers. Think of solar units. Some are linked to the grid while others exist independently. By the middle of the twentieth century, British gasworks consumed about 22 million tons of coal each year.[8]

Figure 5.1 Charging a retort at the gas light factory in Brick Lane, London.

Nineteenth-century dominance of street, factory and residential lighting expanded to include cooking and heating with manufactured gas in the early decades of the twentieth century. Industrial applications and the invention of the gas engine suggested further expansions. Competition from the growing electrical-supply industry for lighting curtailed the manufactured gas industry's growth, however. Although the transition from one form of lighting to the next occurred rapidly during the 1920s, the evidence of increasing competition began in the late 1870s.

With the passage of the first Electric Lighting Acts in 1882, competition from electricity-producing companies challenged the 70-year illumination monopoly of manufactured gas. As electric lights began to replace town gas in 1892, increased industrial applications, cooking stoves, water and household heating with gas took up the slack. The invention of the penny-in-the-slot meter brought working-class households into the marketplace for town gas as well. The invention of a pear-shaped glass mantle by Carl Auer von Welsbach of Germany in 1885 provided another technological and marketing innovation to the town gas industry. His invention-produced light in an asbestos-cotton woven mantle with the cotton dipped in a solution of rare-earth minerals of thorium dioxide and a little cerium dioxide. Each mantle contained a Bunsen burner flame to ignite the cotton, leaving the asbestos and rare-earth minerals intact.[9] When lighted, its glow became incandescent and produced the light of a 50-watt electric light bulb using a minuscule amount of gas similar to that used by a pilot light. By 1905, the incandescent mantle had replaced the old flat-flame burner for street lighting. This innovation helped the gas industry to maintain its prominent position in the illumination business until the 1930s despite competition from the electric light industry.

For much of the nineteenth century, coal gas dominated the manufactured gas industry in Britain and the United States. However, the first urban manufactured gas firm in the United States, the Baltimore Gas-Light Company, initially burned pine tar to produce gas. Other U.S. cities used rosin, British coal, wood and distillate of turpentine. A few years later in the United States, an estimated 3,500 gasworks produced town gas.

With the building of railroad networks in both countries, bringing coal to town gas plants became more affordable. With Pennsylvania mines producing high-quality bituminous coal, the number of gasworks jumped from 30 to 221 in a decade, from 1850 to 1870 to 390 gas plants. With more than 400 in the 1870s located in urban centers throughout the United States, urban residential living changed with the use of coal gas for lighting, heating and cooking despite high prices and competition from electric companies.[10] In Britain and the United States, large coal reserves, advances in organic chemistry and technological innovations help explain the part that manufactured gas played in reshaping urban infrastructures and creating the modern networked city in the nineteenth century.

Since demand outstripped supply, competition for illumination from other energy sources, including coal oil, kerosene and electricity, did not discourage manufactured gas plant owners from lowering their prices or finding markets for coal gas by-products. These by-products included coke, tar, light oil and ammonia, but with a market for coke only, much of what remained became waste. Unlike Britain and the rest of Europe that possessed a coal-based chemical industry for dyes, drugs, explosives and coal tar products, the United States lagged in this growing marketplace.[11]

The global expansion of manufactured gas works

The British design for gasworks became the model for those built in Europe, the United States and around the world. The British also possessed the greatest density of gasworks with 2,327 locations. The seamless transfer of manufactured gas technology from Britain to the United States, given their common language and culture, grew rapidly by the mid-nineteenth century, with 50 cities using gas lighting for streets, commercial and elite housing. The larger urban population continued to use whale oil and tallow candles for lighting until prices became prohibitive. Whale oil, typically priced at US$0.80 a gallon at mid-century, saw spikes in prices as overhunting depleted the global whale population. For modest and low-income families, tallow candles that burned for seven hours cost US$.025 each. New York City, Philadelphia, Baltimore, Cincinnati and Chicago led the country in the number of consumers burning tallow candles.[12]

Paris and Lyon, France, maintained their own gasworks, with 208 others located throughout the country. In Berlin, Germany, 10 gasworks produced town gas, and by the second decade of the twentieth century, 263 were located in central towns and cities. The Dutch built a gasworks in Rotterdam in 1825. In the years that followed, 122 additional gasworks produced town gas. In 1855, the atmospheric gas burner invented by the German chemist, Robert Wilhelm von Bunsen, provided the intense heat needed to establish a market for manufactured gas as a fuel for industrial heating. In 1876, Nikolaus Otto invented an alternative to steam-powered engines with the four-stroke gasoline engine. British and German manufacturers begin building separate producer gas plants at their factories. In Belgium, scientists perfected by-product coke ovens, recovering light oils, sulfur and ammonia as by-products. Gas, however, remained the primary by-product of coking in 148 other plants.

Between 1835 and 1871, Imperial Russia built 31 gasworks using British industrial designs. In Switzerland, 63 gasworks delivered manufactured gas to 220 Swiss parishes by 1912. Australia used its abundant coal reserves, lime and iron ore to produce gas in 141 plants. New Zealand's proximity to Australia allowed local entrepreneurs to purchase coal from Newcastle, New South Wales, for its gas works. Japan built 41 gasworks using British models with much of the growth occurring after its victory in the Russo-Japanese War of 1905. For much of the world, gasworks in double and single numbers operated to produce town gas. India, China, Brazil, Argentina, Italy, Spain, Sweden, Denmark, Mexico and Chile maintained plants in double numbers. The countries that remained operated fewer than 10 gasworks.[13]

The inventive process

As the market for town gas spread in the nineteenth century, from illumination of streets, factories and elite households to cooking and heating, inventors refined the production process. In 1873, Thaddeus S.C. Lowe invented water gas as a more efficient way to produce manufactured gas. His method involved the passage of high-pressure steam over burning coal or coke, producing carbon monoxide and pure hydrogen. Further processing by cooling and scrubbing removed impurities and passed it through water vapor, producing pure hydrogen gas (water gas). Initially, water gas burned poorly when compared to coal gas. In 1875, however, Lowe discovered that by spraying water gas with liquid hydrocarbons, he produced "carbureted water gas." With a higher-BTU rating of between 500 and 600 than its predecessors, it gained the majority of users. In 1914, the manufactured gas industry produced 44.2 percent water gas, 36.4 percent as a mixture of coal and water gas and 5.2 percent coal gas.[14]

The inventor Thomas Alva Edison began working in his laboratory in Menlo Park, New Jersey, determined to replace town gas's foul odor and, more important, its propensity for fires and explosions with electric lighting. Developing an electric light and a system of transmission to households and businesses became the focus of Edison's mission. On October 15, 1878, he incorporated the Edison Electric Light Company to protect his patents on electrical devices. He hoped to string electric cables through existing gas pipes, but gas companies refused to grant a competitor such easy access to its infrastructure. Under the name of the Edison Electric Illuminating Company, he built an electricity-generating company on 257 Pearl Street in New York City that began operating on September 4, 1882. By incorporating his utility under the city's gas laws, he was authorized to lay electric wires under the city streets.

Edison's invention spread quickly from city to city, replacing town gas as a major source of lighting. With the development of the tungsten filament, the cost of electric lighting competed favorably with the cost of town gas. To maintain market share, town gas companies consolidated, lowered prices and innovated with the new Welsbach gas mantle. By 1910, however, electric lighting had replaced gas lighting in most cities. To survive, the manufactured gas industry diversified its portfolio of services by maximizing its expansion into heating and cooking. Advertisements praised the benefits for gas stoves, furnaces and gas appliances for household and commercial use. Abandoning the lighting market and focusing on cooking and heating sustained the manufactured gas industry into the mid-twentieth century.[15]

The environmental legacy of the manufactured gas industry

With the closing of the last manufactured gas plants in the United States in 1966, manufactured gas left a toxic brew of discharged chemicals in the grounds surrounding abandoned plant sites. Tars made up of 300 to 500 different compounds are toxic to humans, animals and plants. Some compounds were carcinogenic. Ammonia, cyanide, sulfur, and heavy metals, including arsenic, saturated the soil caused by drips and spills. These contaminants caused by carelessness, accidents, leaking storage tanks and normal manufacturing processes left a legacy of polluted soil, much of which remained invisible for decades and long after the termination of manufactured gas processing at an estimated 52,000 or more plants nationwide.[16]

Although citizens sued gas plant owners for the noxious odors fouling the air in neighborhoods affected by the manufacturing process, courts and municipal and state boards of health responded with statutes favorable to plaintiffs but with weak enforcement provisions. As historian Joel Tarr has pointed out after surveying the technical literature from approximately 1880 to 1919, only four published articles dealt with issues of waste disposal and pollution.[17] Focusing on the atmosphere with ordinances to control smoke emissions and unseemly odors missed conditions found in place. As historian Peter Thorshem has pointed out, in Britain "conditions inside gas works and coke plants were harsh, unhealthy, and sometimes deadly."[18] Coal carbonization inflicted a range of injuries, from severe burns to severed limbs, on workers. For those who survived the daily hardships of coming into contact over five years with the products of coal carbonization suffered lung cancer 10 times greater than the general population.

While carbonization filled the atmosphere with massive amounts of black smoke, the land absorbed a punishment that would take decades and possibly centuries to ameliorate. Smoke and terrible stenches do not begin to describe the contamination caused by gasworks. As early as the 1820s in Britain, residents complained about the damage to the land in the form of wilted plants, stunted tree growth, soiled clothing, tarnished brass and copper and respiratory illness.

Rivers and streams in Britain became sinks for by-products of coal gasification that possessed no commercial value. Besides leaks and drips, plant managers allowed these derivatives to drain into the nearest waterway. Poorly enforced laws prohibiting such activity could not keep up with the volume of pollutants dumped on the soil and into the water. As early as 1821, London's Thames River became contaminated with dead fish and eels. Unlined holding ponds, dug to prevent by-products from flowing into waterways, seeped into the groundwater polluting well water. The volume of discarded waste overflowed in streams and rivers. Discarded coal tar with its numerous toxic compounds upended the delicate ecological balance of the land.

In households, burning manufactured gas for lighting proved to be more effective than whale oil or candles and cleaner than coal-fired lighting, but gas lights emitted a sharp odor, deposited soot on furniture and fixtures and stunted the growth of household plants, to say nothing of its effect on plant, animal and human respiration. Efforts to further purify gas at the works usually required processes that displaced odors from the household to the plant site. In the 1870s, town gas operators, concerned that coal gas smelled, created soot, and burned with a dull light, produced water gas. This was achieved by forcing steam across the beds of coke in a retort. Oil vapors condensed to produce a brighter light with reduced odor and less soot.

Not until the passage of legislation that created the Environmental Protection Agency (EPA) in 1970 did a U.S. government agency identify polycyclical aromatic hydrocarbons (PAHs), coal-tar emissions and a by-product of coke ovens as dangerous air pollutants. By enforcing new clean air regulations, the by-product coke industry in the United States ended by 1990. The EPA uncovered the damage done to air and groundwater by the defunct manufactured gas industry. Since the utility industry owned many of these defunct gasworks sites, the sites came under the jurisdiction of the federal Superfund Act of 1980. The EPA identified such sites for remediation, with many state environmental agencies using federal funds for that purpose. Many sites have escaped remediation on the grounds that they pose no immediate threat to humans or to the environment. This designation remains a contested conclusion to many environmentalists who claim that the operational history of these sites has not been given adequate attention in escaping a superfund designation.[19]

The history and development of the natural gas industry

Natural gas, composed of mostly methane is the cleanest fossil fuel. Like oil, the organisms that lived in the oceans millions of years ago provided the substance that eventually became oil and gas. While some organisms consumed others, those that remained became buried in overlaying ocean sediments. Increasing sediments subjected the surviving organic matter to steadily increasing temperatures and pressures. The rising temperatures reflected the heat from Earth's mantle, referred to as the geothermal gradient. At greater depths of 7 kilometers, or more than 4.5 miles, temperatures greater than 230 degrees Celsius, or 446 degrees Fahrenheit, produce low molecular-weight hydrocarbons that make the formation of natural gas possible. Much of it migrates to the surface of the oceans and is released into the atmosphere as methane. Some of it is captured by impermeable rock or trapped shale. Millions of years later, modern civilization hungry for this stored energy tapped into these reserves to power industry and households.[20]

Small amounts of natural gas escaped from Earth's mantle and surface and its oceans for millions of years. Leaking gas often caught fire ignited by the sparks of falling rocks and by lightning strikes. Ancient Greek civilization, more than 3,000 years ago, erected a temple to Apollo, its god of truth and light, on the sight of escaping gas that would provide fuel for its perpetual flame. Chinese civilization 2,500 years ago harvested natural gas and transported it by a pipeline made of bamboo stems to a site for boiling brine to collect salt and for transporting potable water. The first commercial use of natural gas in occurred in the Renaissance city of Genoa, Italy, that lighted its city streets with natural gas transported from wells in nearby Andiamo, Parma. Centuries would pass before natural gas fulfilled a similar function in the remaining Western world.[21]

In the United States, natural gas entered an energy regime dominated by coal, manufactured gas and oil in 1821. By accident, boys in Fredonia, New York, about 40 miles from Buffalo threw burning sticks across a creek and watched as flames shot up from the ground. Their burning sticks hit leaking gas. The boys' family capped the leak with a tin roof and ran a pipe to their house to light their home and cook their food. In 1840, wildcatters discovered natural gas in Centerville, Butler County, Pennsylvania. Entrepreneurs used it to evaporate brine to produce salt in much the same way as the Chinese produced salt millennia ago. Not until 1858 did the Fredonia Gas Light Company tap local gas wells to deliver natural gas to commercial and residential buildings for lighting.[22]

Making the transition from manufactured to natural gas faced obstacles from independent distributors. Opposition surfaced when these distributors tried to compete with existing utilities for control of natural gas and the transition away from coal gas. Manufactured gas production depended on the size and distribution network of its producers and consumers. That network included a robust array of coal reserves, coal-gas plants, pipelines, local distribution centers and appliances for households and commercial enterprises. Collectively, they represented a large capital investment by the producers. With so much invested in coal gas, producers were reluctant to switch to a cheaper, more efficient natural product that burned at a higher BTU intensity.[23]

Where public utilities controlled the infrastructure to distribute manufactured gas, the transition to natural gas proceeded without obstacles. New pipelines to introduce natural gas to non-elite consumers proceeded unevenly. Before natural gas entered the energy stream for consumers, it required treatment to remove impurities. Since it came from different reservoirs deep into Earth's mantle, it was unevenly distributed in the environment. Unlike manufactured gas produced in retorts, natural gas required discovery, purification and distribution through new and existing pipelines. Some natural gas contained high concentrations of sulfurous compounds. Once purified and ready for transmission to consumers, natural gas contained mostly methane, with trace amounts of ethane, propane, butane, and with a small amount of an odorous substance injected to detect leaks.

Natural gas consumption spread more rapidly after the discovery of oil by Edwin L. Drake in Titusville, Pennsylvania, in 1859. Where there was oil, natural gas pressurized oil wells, driving the crude to the surface. With no known immediate use for natural gas, oil well operators allowed the gas to escape into the atmosphere. Within a short time, however, operators recognized the high-energy value of natural gas. Compared to manufactured gas with a BTU value of approximately 500, natural gas, regardless of the reservoir from which it emerged, maintained an energy value of 1,020 BTU. Much of it was used in oil fields and in refineries. Recognizing its energy value, local industries began using it for heat and lighting. Beginning as early as 1860 and lasting until the 1920s, natural gas reserves in Pennsylvania, West Virginia, New York and Ohio remained the largest in the United States.[24]

As late as the 1870s, Rochester, New York, Natural Gas Light Company used wooden pipelines 25 miles long to deliver natural gas from the well to consumers. Made of Canadian white pine, using 8-foot-long logs, planed to a 12.5-inch exterior diameter and bored to an 8-inch interior diameter, they represented the best that technology offered for transmitting natural gas at the time. For 25 years, a wooden pipeline delivered gas from a well in Westfield, New York, to a lighthouse on Lake Erie. Yet, a similar pipeline delivering gas to Rochester lasted only a few years. Rotting pipelines and leaks required a transition to a more reliable 2-inch wrought-iron pipe in 1872. Its diameter and its low pressure made it suitable only for short-distant transmission.[25] At the time, pipeline technology served as a barrier to production and transmission. Those in existence served local markets inefficiently and dangerously. Gas explosions caused the loss of life, injuries and property damage.

Pittsburgh and natural gas

Pittsburgh's City Council in 1882 gave gas companies the right to dig up the city's streets to lay gas pipelines. "But since the companies had little experience in this regard, their work was substantially flawed, with improper piping for mains, numerous leaks around joints and valves and inadequate allowance for wintertime temperatures."[26] Despite these limitations, in the late 1880s the region possessed 107 producing gas wells and 500 miles of pipelines. Of these, 232 miles connected residences and local industry, manufacturing iron, steel, glass and chemical products.[27]

Although coal, oil and manufactured gas dominated industrial and domestic needs for energy, the discovery of natural gas created enthusiasm for the fuel's energy density. The discovery of natural gas that became known as the McKeesport, Pennsylvania, gas boom represented the extent of this enthusiasm. In 1919,

> many residents signed leases for drilling on their land. They bought and sold gas company stock on street corners and in barbershops transformed into brokerage houses. Houses were torn down and front yards uprooted. Hundreds, and even thousands, of wells were supposedly drilled in the 9-square-mile area.[28]

A year later, the Pittsburgh district's biggest natural gas boom crashed, with companies and citizen investors becoming heavy losers.

The expansion of the natural gas market

The boom/bust cycle that the Pittsburgh district experienced was preceded by a similar one decades before in Indiana as the state made the conversion from manufactured gas to natural gas in statewide street lighting. The transition began in 1886, but by 1907, state reserves of natural gas were depleted, forcing a return to manufactured gas.[29] Regionally, however, pipeline technology evolved rapidly as the demand grew for natural gas in cities across the American Midwest.

The first high-pressure pipeline was laid in 1891, bringing gas from fields in Indiana to Chicago. At 120 miles in length, it carried gas at a pressure of 525 pounds per square inch. Welding the joints to prevent gas from leaking under high pressure was achieved with oxygen-acetylene welding first used in 1911. In 1925, the invention of seamless electrically welded pipe ensured relatively smooth insides. Without a smooth interior, gas would be slowed as it moved from well to its destination. Only then did the gathering and transmission of gas become profitable across greater distances. This technological development signified the beginning of the end for manufactured gas.

The Great War's (World War I's) demand for coal in the unseasonably cold winters of 1916–1917 placed the manufactured gas industry in a precarious position. The need for a coal-tar substance (benzole) during the war as an illuminant convinced the British War Office that the energy produced from manufactured gas should be measured in caloric value (the amount of energy produced by complete combustion) rather than candlepower. In addition, the new standards made it profitable to recover toluol (a carbon-based solvent), an essential chemical in the production of explosives and other light oils for the war effort. Britain's Gas Regulation Act of 1920 began charging consumers by the therm equal to 100,000 BTUs to guarantee its value, purity and pressure. A therm, taken from the Greek word for *to heat*, is 22,500 kilocalories of heat, or 96.7 cubic feet of gas. By the end of the decade, town gas sales in Britain reached 230 billion cubic feet annually.[30]

In the United States, the federal government placed emergency controls on the supply and distribution of coal during the Great War, the nation's major fuel source. Natural gas replaced fuel oil and coal in industries engaged in manufacturing military equipment. During the war, the United States Fuel Administration intervened directly in the productive capacity of the natural gas industry by requiring it to also produce toluol. The Consolidated Gas Company in New York produced charcoal, a necessary filter in gas masks to protect soldiers fighting in France from the chemical warfare being waged there.[31] With the end of the war, controls ended, but 3,600 coal mines strikes in 1919 involving 4 million miners brought back government controls temporarily. By 1920, competition from oil began to erode coal's dominant position as an energy source, with natural gas achieving renewed interest because of its wartime importance.

The use of natural gas for heating occurred slowly because early appliances lacked flues to exhaust carbon monoxide. The invention of the firebrick back, asbestos fiber and woven wire to radiate heat from an underlying gas fire accelerated the development of safe heating appliances. The instantaneous gas-powered water heater became available in 1890, transforming thousands of cold-water households by providing a continuous hot-water supply. Improvements in gas stoves led to the replacement of wood- and coal-fired black cast-iron cookers with enamel cast iron in 1913 and enamel pressed-sheet steel a few years later. Manual controls for ovens experienced similar innovations with the introduction of the thermostat oven in 1923.[32]

By the second decade of the twentieth century, the discovery of two large gas fields in the American Southwest shifted focus from Appalachia to the Panhandle field in North Texas and the Hugoton field that bordered Kansas, Oklahoma and Texas. Much like the experience of drilling for oil in western Pennsylvania in the later decades of the nineteenth century, drillers viewed natural gas as a waste product and allowed more than a billion cubic feet a day to escape into the atmosphere.[33] Recognizing its energy value, commercial interests explored ways to make it a saleable commodity.

During 1928–1929, pipelines connected the Panhandle field to Kansas City, Missouri, and Denver, Colorado. This 500-mile interstate pipeline served Pueblo and Colorado Springs as well. Reaching Denver made it the hub for natural gas distribution, including Idaho, Wyoming, Montana, Utah and Nevada. Most of the growth would take place in the West and South, in California, Texas, Oklahoma and Louisiana. By 1931, however, the north Texas field 1,000 miles away from Chicago delivered natural gas to the city through a 24-inch welded steel line. In the same year the nation's capital began receiving natural gas from the same field and mixing it with manufactured gas. It stopped using the mixture in 1946, phasing out town gas.[34]

In Colorado, deliveries quadrupled to domestic customers and doubled to industry. In Denver, under the leadership of Henry L. Doherty, the combined gas and electric utility Cities Services Company began an intensive marketing campaign by selling new gas and electric appliances to residential consumers. Focusing on a set of common concerns to women, including cleanliness, comfort and convenience, Doherty's salespersons promoted the virtues of electric lighting, electric irons, gas stoves and gas-fired hot-water heaters. By contrast, meals cooked over a coal-fired stove required extensive cleanup of soot from the stoves, windowsills and upholstery. After 1900, architects and builders extolled the virtues of gas and electric appliances. Fifteen years later, the curriculum at vocational schools provided instruction about their installation and repair. Young women learned to cook with gas and to use electric sewing machines in home economics classes. Similar developments, extolling the benefits of these new gas and electric appliances to a growing population of consumers, took place across the country.

The Great Depression (1929–1938) disrupted the expansion of the natural gas industry. Without capital investment, no new long-distance pipelines became operational. Oversupply of gas plagued the industry in the Southwest, and consumers lacked the power to borrow money or the income to purchase energy. Oversupply in the Southwest led to shortages in the Northeast. Although coal faced increasing competition from oil and natural gas, during the 1930s consumers came to rely on locally produced manufactured gas for cooking and heating in the Northeast. As the Great Depression began, Pacific Gas & Electric, tapping gas wells in Southern California, delivered natural gas to San Francisco and Oakland. Without a market for their gas reserves, Texas, Oklahoma, Kansas and Louisiana gas companies faced bankruptcy. Texas oilmen determined to protect their petroleum business returned to the wasteful practice of venting "trillions of cubic feet of unmarketable 'waste gas' into the atmosphere."[35]

During World War II (1939–1945), government-financed pipelines measuring 24 inches and 12 inches in diameter delivered crude oil from East Texas to New Jersey refineries. Refined into gasoline, diesel and aviation fuel, it provided the energy needed to keep planes, ships, tanks and other military vehicles engaged throughout the war. German submarine attacks on oil tankers carrying oil from the Gulf Coast to New Jersey refineries accelerated the construction of pipelines. With victory in 1945 and a return to a peacetime economy, 1,000-mile-plus pipelines to New Jersey were sold to private companies in 1947 for the purpose of transmitting natural gas. A privately built 1,840-mile gas pipeline from Texas to New York City became operational 1951. Natural gas mixed with manufactured gas became part of the energy equation in New England a year later.[36] After World War II, forced-air furnaces burning natural gas in the Midwest and West replaced coal stoves and furnaces.[37] In the Northeast, oil delivered by pipeline replaced coal, but the transition to abundant and affordable natural gas would await the conversion of existing pipelines and the introduction of new ones. The use of mixed gas and oil and gas mixture represented a temporary transition only to be replaced quickly by natural gas. Nationwide, 80 percent of consumers depended on natural gas while only 20 percent continued using manufactured gas.

The modern natural gas pipeline system

Rapid expansion of the natural gas industry led to the creation of cartels involved in consolidating control over natural gas reserves and distribution networks. In response, the Federal Trade Commission launched an investigation to study waste, unregulated distribution, monopolistic control of gas production and financial fraud. The commission's findings led to federal regulatory statutes intended to discourage abuses and standardize business practices. Among these statutes were the passage of the Securities Act and Securities Exchange Act of 1933 and 1934, respectively; the Public Utility Holding Company Act; and the Federal Power Act (1935) and the Natural Gas Act (1938).[38]

In the decades following World War II, and after public policies to regulate the industry had become law, the natural gas industry expanded by the discovery of new gas reserves and pipeline construction. As noted earlier, technological developments including higher tensile strength for pipes, advanced gas pressure welding, and the invention of high-pressure gas compressors to move gas through the pipelines allowed utilities to move high volumes of natural gas from fields to households and industries.

The expansion of the natural gas distribution network required the conversion of appliances dedicated to low-BTU manufactured gas to natural gas. As the country's largest city in terms of population, the New York episode was noteworthy for the magnitude of the conversion process. First, beginning in 1949 Transcontinental, a pipeline company built a 30-inch 1,000-mile pipeline extending from the Rio Grande Valley in Texas to Manhattan. Completed in 1951 at a cost of US$233 million, it delivered natural gas to the country's richest city. Five major utility companies shared the metropolitan market and the costs of underground pipes with Consolidated Edison (more than a million customers) and Brooklyn Union (350,000 customers) sharing the city's market.

Short-term conversion costs required meeting New York's stringent safety regulations of running pipelines under city streets accustomed to heavy vehicular and pedestrian traffic. For all five companies, the cost of a 262,830-foot pipeline reached US$14 million. By contrast, outdated manufactured gas plants, including its large storage gasometers owned by Con Edison, originally represented a US$250 million investment. Average consumers owned two manufactured gas appliances, in all probability oven ranges, water heaters or gas refrigerators. Experienced workers visited each household to adjust appliances used previously to burn low-BTU manufactured gas to the cleaner, more efficient, and cheaper higher-BTU natural gas. In some cases, manufactured gas employees laid off from their plants became skilled in making the household conversions.

By 1956, Con Edison accomplished the conversion at a cost of approximately US$36 million. The company delayed decommissioning its manufactured gas plants, however, thinking that they would serve as a backup for customers if the natural gas pipelines became disabled. Their use, however, ranged from rarely to not at all. For Brooklyn Union's customers who owned more than 2 million gas appliances, the conversion process began in March 1952 and was completed in six months. For its industrial customers who used gas furnaces, the process was completed by the end of the year.

Once pipelines reached New England in 1953, a similar conversion process took hold, and by 1959 the manufactured gas industry's 100-year domination reached a historic low in terms of customer usage. Its standing as a significant energy industry in the country ended as the natural gas industry not only replaced manufactured gas but also added new customers, becoming the country's fastest-growing fossil fuel. From 1950 to 2015, the United States consumed almost 5 million cubic feet to more than 27 million cubic feet, respectively.[39]

The growth in natural gas usage proved to be sporadic, however, caused by the steady growth of hydropower and nuclear energy, to generate electricity for American households. The price controls imposed during the Great Depression and the tightly regulated natural gas markets dissuaded developers from investing in domestic gas exploration. The oil crisis of 1973 caused by the decision by the Organization of Petroleum Exporting Countries (OPEC) to curtail exports increased the price of petroleum. Federal regulators convinced that domestic reserves of natural gas were limited depressed consumer demand for affordable energy. With this unproven assertion, Congress passed the Power Plant and Industrial Fuel Use Act (FUA) in 1978 that prohibited the construction of new gas-fired power plants. Between 1978 and 1987 when FUA was repealed, the United States added 172 gigawatts of power generation, 81 of which was coal-powered, with nuclear power generating half of the remainder.[40]

At the same time, Congress deregulated the industry with the passage of the National Energy Act (1978). All gas in production before 1978 would remain regulated; all newly discovered on-shore gas would be priced at US$1.75 per Mcf (1,000 cubic feet of natural gas) and deregulated thereafter. That deregulation would unleash a great demand for natural gas proved to be wrong. Conservation measures promoted during the first and second oil crises for residential customers convinced gas customers to conserve more and use less. As a result, natural gas energy consumption declined in the 1980s compared to the 1970s and did not increase again until the 1990s.[41]

By then, highly efficient and relatively inexpensive gas turbines and technologies to explore deep water, offshore gas reserves removed perceptions about limits. Between 1989 and 2009, the United States added 306 gigawatts of power generation, with 88 percent of it in gas-fired capacity and only 4 percent as coal-fired. Another cycle of "feast and famine" followed in which capacity outstripped demand. Once demand caught up, conventional supplies declined, while offshore supplies remained expensive. Rising prices for gas and the belief in long-term shortages accelerated the development of a liquefied natural gas (LNG) infrastructure, now largely unused given the shale gas revolution of the last years.[42]

In Britain, the Great Depression of the 1930s and the increased use of electricity led to the nationalization of the manufactured gas industry in May 1949. Henceforth, 12 Area Boards reporting to the minister of fuel and power controlled the industry. A similar fate happened to the coal industry. By 1957, as many as five of six households used manufactured gas for cooking, heating and lighting.

The conversion from manufactured gas to natural gas

The conversion narrative began with the discovery of natural gas in the North Sea's West Sole gas field in United Kingdom waters off the southern coast of England in November 1965. Despite the known hardships of drilling and transporting gas in the unpredictable weather of the North Sea, it surprised both experts and observers with its ferocity. George Brown of the energy service company Brown & Root noted, "Some months we were able to work only three or four days. Any sea can get rough, but the North Sea lays it on. During one period we expected to lay forty miles of pipeline. We laid only nine."[43] The company met the challenge by laying a 45-mile, 16-inch pipeline at a depth of 105 feet to onshore gas facilities of the British Gas Council at Easington, East Yorkshire. The completion of this pipeline led to the company laying a 24-inch pipeline for almost 2 miles under the Humber River, connecting it to the Gas Council's existing grid used for the transmission of manufactured coal gas. In 1966, the discovery of three additional gas fields suggested the beginning of an era of energy independence for Britain.

Unlike the decentralized gas industry with hundreds of utilities in the United States, Great Britain had nationalized its industry following World War II. With these gas field discoveries, the Gas Council's chairman, Sir Henry Jones, announced the decision to convert the country from manufactured coal gas to natural gas. Planning for onshore distribution terminals and the development of a high-pressure transmission system commenced immediately. The conversion took about 10 years and involved 13 million customers at a cost of 1,000 million pounds. Much like the process in the United States, the council trained laborers to adjust household appliances to burn a higher-BTU fuel. Unlike some U.S. cities that mixed manufactured with natural gas and converted later to natural gas solely, the centralized British system avoided this costly intermediate step. Compared to the experience in the United States that took 30 to 40 years, the 10-year conversion in Britain compared favorably to it and to the Dutch conversion of 2 million customers in 6 years and the Japanese with 5 million customers in 12 years.[44]

Origins of the shale gas revolution

An oilman named George Mitchell, owner of Mitchell Energy & Development Corp., invested US$6 million over 10 years trying to extract oil and gas from the Barnett Shale of North Texas. In 1998, with the high price of oil, he succeeded in making his investment profitable by using a variety of chemicals and propellants to recover enough hydrocarbons from the wellhead known as S.H. Griffin #4. At the time of his discovery, shale gas and oil represented only 2 percent of the nation's domestic supply. In 2002, Devon Energy Corp. purchased Mitchell Energy for US$3.5 billion in cash and stock. Devon Energy added horizontal drilling to its repertoire and, along with other energy companies, greatly expanded their ability to profitably drill for shale gas. By 2008, 25 percent of the domestic supply came from shale. In 2013, shale gas accounted for 37 percent and combined with light oil 50 percent of the total domestic supply. By 2035, estimates have placed shale gas and oil at 80 percent.[45] The Marcellus Shale (black, low-density, carbonaceous shale from the Middle Devonian Period [393–382 mya]) in New York and Pennsylvania is the largest gas field in the United States, with sizable gas fields located in Ohio, West Virginia, North Dakota, Arkansas, Louisiana, Oklahoma and Texas.

In the past decade, the United States has overtaken Russia as the world's largest producer of natural gas and is expected to exceed Saudi Arabia's production of oil by 2020. How has this happened? Energy companies have tapped shale oil and gas reserves by developing hydraulic technologies to fracture shale deposits 6,000 to 10,000 feet below the Earth's surface. Fracturing occurs by pumping water under high pressure into the shale rock, causing it to crack and releasing both natural gas and light oil. Combined with sand to prop open the fractures, called "proppant," the flow of water prevents the cracks from resealing, allowing a continuous flow of gaseous and fluid fossil fuels. With sand as the primary proppant, chemical agents also play a vital role by keeping the sand in place and in preventing bacteria from degrading the fuel. The second technology employed to capture shale gas and oil is horizontal drilling.

After drilling a curved borehole to a depth of 1 mile or more, operators turn the drill sideways to increase contact with the layer of shale that contains gas and oil. Steel and concrete line the borehole. When reaching the shale rock, a tube-shaped device called a perforator, a gun-like mechanism filled with bullets, is lowered into the borehole. When triggered, the bullets penetrate the borehole embedded in the shale. Thousands of gallons of water, pressurized at 20 times that of the pressure in a garden hose mixed with sand, exit the bullet holes in the well casing fracturing the shale. While drillers used the technique of fracturing to release oil from wells in the 1940s, it was used to release gas from coal beds in the 1980s. In fact, explosives released oil from wells in the nineteenth century.[46]

Among the many complaints surrounding hydraulic fracturing, or "fracking," include contamination of fresh drinking water, earthquakes, oil spills and methane leaks. One complaint seldom receives much scrutiny. The huge amounts of "frac sand" keep the fissures in the shale open. The proppant ensures a steady flow of gas and oil from the shale. At the height of the shale gas boom in 2014, the frac sand industry ballooned to US$4.5 billion. The decline of oil and gas prices followed, depressing the frac sand industry. In 2016, it declined to become a US$2 billion industry. Despite fluctuations in the prices of oil and natural gas caused by supply exceeding demand, the mining of frac sand degrades the land and imposes unintended environmental costs.

Frac sand exists in a number of locations, including, Texas, Kansas, Arkansas and Oklahoma. In none of these states does it exist beneath prime farmland. In the upper Midwest, however, Illinois, Wisconsin, and Minnesota possess some the world's richest farmland. It sits atop a highly valued source of fine silica called St. Peter Sandstone. Since the beginning of the shale gas revolution, industrial sand facilities that include mines, processing plants and railheads had purchased 3,100 acres of the best farmland in Illinois from 2005 to 2014. By the end of 2015 as many as 129 sand plants operated in these upper Midwest states.

Stripped of the rich soil that formerly produced food crops for people, it now became known as "overburden." Piled as high as 30 feet of topsoil and subsoil, it exposed the frac sand, with a high silica content and uniform grain size and strength. When functioning at full strength, the industrial sand facilities become 24-hour operations, with mine blasting releasing the sand. Hundreds of diesel trucks strain the capacity of rural roads to withstand the weight of vehicles and their loads of sand. With trucks dropping some of the sand along the way, airborne silica posed a health hazard for workers and residents of nearby farms and rural communities. Silica is a known carcinogen that causes lung disease, including silicosis. This threat to human health and the millions of gallons of groundwater used to operate the mines compromised the vitality of rural communities in the Midwest.[47]

Fracking a gas well requires taking about 4 million gallons of water treated with chemicals, some of which are proprietary secrets. This combination, along with 3 million pounds of silica sand, is forced into a 1- or 2-mile-long well through the shale. Cracking the shale releases oil and natural gas that percolates to the surface along with the fracking fluid, a toxic brew of bacteria, heavy metals and proprietary chemicals that contaminate the surface and pollute the air. Some of the fracking fluid remains in the ground, combining with methane that can migrate into aquifers, streams and drinking water wells, compromising human health. "Mouth ulcers, severe abdominal pain, nausea, swollen lymph nodes and dizziness" and dead pets plague residents of rural communities where drilling companies release fracked oil and natural gas.[48]

Hydraulic fracturing has unlocked millions of barrels of tight (light) oil. From 100,000 barrels a day in 2003, the amount skyrocketed a decade later to 2 million barrels per day. With production expected to reach 11.1 million barrels a day by 2020, the United States will pass Saudi Arabia's daily output. The current

Figure 5.2 A diagram of fracturing shale in order to release oil and natural gas.

and future oil output of the Bakken Shale in North Dakota succinctly defines the magnitude of this transition from energy dependence to independence. As a 25,000-square-mile reserve of embedded oil, this shale deposit possesses an estimated 11 billion barrels of recoverable oil using current hydraulic fracturing technology. As the technology improves, the estimates of recoverable oil rise to as much as 30 billion barrels. North Dakota passed California and Alaska in oil production in 2012 to become second only to Texas.[49]

The shale revolution in the United States has an impact on its economic system and its geopolitical standing in the world. As a capital-intensive industry, its investments may reach as high as US$5.1 trillion in 2035. It supports high-wage drillers who create an estimated three to four other jobs in machinery, financial services and geology. Added jobs reduce unemployment and generate higher levels of consumption for goods and services, elevating the gross domestic product. Increased production of domestic gas reduces prices for heating and electricity in residential and commercial buildings. They represent 50 percent of the nation's energy bill. When prices fall below production costs, it ripples through the economy, causing layoffs of workers and reductions in drilling. The "feast and famine" cycle creates economic disequilibrium.

With almost one-third of the energy consumption in the United States coming from transportation, the price of oil is set on the world market, not on domestic demands and supplies. An oil glut internationally depresses prices and increased production in the United States alone will not alter pricing as much as domestic natural gas production. Natural gas is much cheaper than petroleum. Gasoline-powered cars, trucks and buses pollute the atmosphere and cost consumers billions in operating expenses. Currently, the technology exists to power motor vehicles with natural gas. Many cities now purchase natural gas–powered buses and trucks, and LNG offers motorists the potential to power their electric and hybrid automobiles. If the significant cost differential between natural gas and gasoline persists, manufacturers and innovators will find a way to power cars and trucks with natural gas.[50]

The geopolitical implication of becoming energy independent reduces the vulnerability of the United States to many of the world's leading energy suppliers. For oil exporters, Saudi Arabia, Russia, Iran, the United Arab Emirates, Norway, Iraq, Angola and Nigeria lead the way. For natural gas suppliers (not shale gas), Iran, Qatar, and Russia control 70 percent of the world's known reserves. Some, but not all, of these countries maintain adversarial relationships with the United States. Saudi Arabia, ostensibly an ally, supports an Islamic sect, Wahhabism, that promotes Jihad against the West. Achieving energy independence strengthens the geopolitical position of the United States in dealing with other energy-producing countries. In the long run, it may allow the United States to reduce its yearly budget of US$60 to US$80 billion to protect the flow of oil in the sea lanes of the Middle East. In 2005, the United States imported 60.3 percent of its oil. By 2013, the nation cut this figure in half. The International Energy Agency projects that the United States will become 97 percent energy self-sufficient by 2035.[51]

Global shale gas reserves

Ten countries possess 79 percent of the world's shale gas reserves. China's known reserves place it at the top of the list, with an estimated 36.1 trillion cubic meters of recoverable shale gas. However, China lacks the natural gas infrastructure, including pipelines and roads, to exploit the fuel's full potential. Also, China's shale is buried twice as deep as that in the U.S. and its clay composition makes it more difficult to frac. However, the country's 12th five-year plan presented in 2011 seeks to reduce carbon dioxide emissions by 17 percent. As recently as 2011, however, China consumed 3.48 billion tons of standard coal equivalents annually with coal accounting for 68.4 percent, oil for 18.6 percent and natural gas for 5 percent.[52] Aggressive goals of reducing China's dependence on coal and reversing its carbon dioxide emissions will depend on massive investments in shale gas exploration and the inclusion of Sino-foreign joint ventures. By mid-2012, China signed its first contract for drilling and commercial production of shale gas. Whether shale gas is the "game-changer" that transforms China's energy system is anticipated but not known presently.

Argentina and Algeria place second and third. As one of the world's most robust economies, the United States is the world's largest energy consumer and places fourth with an estimated 18.8 trillion cubic meters. Canada, Mexico and South Africa follow the United States while Russia, with its vast recoverable deposits of all fossil fuels, places ninth with an estimated 8.1 trillion cubic meters. Tenth place belongs to Brazil, with 6.9 trillion cubic meters. In all, these 10 countries possess 163.1 trillion cubic meters of shale gas resources globally, with the rest of the world's countries holding an estimated 43.5 trillion cubic meters.[53]

The discovery of the world's second-largest shale formation in Vaca Muerta located in southwestern Argentina would make the country a main natural gas exporter in the coming decades. Building new infrastructure and improving existing pipelines will require a large capital investment. Toward that end, Shell Argentina and other international energy firms increased capital expenditures by US$500 million in 2014. Algeria also finds itself in a favorable export position. If developed, its known shale reserves could supply the entire European Union for a decade with much-needed fuel. Canada's recoverable shale gas is located in most of its provinces. As recently as 2013, however, the country witnessed anti–shale gas demonstrations by activists.

With the sixth-largest recoverable shale gas reserves, Mexico's minimal investment in development will result in natural gas shortages as the country's population become high-end consumers. Its large oil reserves may not be sufficient to close the gap between energy supplies and consumer demand. In contrast, the Republic of South Africa, with vast reserves in the semiarid Karoo region, issued drilling licenses and regulations to international energy firms in 2013. Russia remains committed to developing its vast natural gas reserves and avoiding shale-gas exploration. Brazil, profiled in Chapter 7, generates 80 percent of its electrical power needs from hydropower. However, drought conditions in the last decade have pushed dam levels to low levels. Currently, it imports natural gas. Shale gas development may make it an exporter in future decades.[54]

The costs and benefits of shale gas production

Since 2012, vast reserves of shale gas have grown from 4 percent to more than 25 percent of the U.S. supply of natural gas. These new reserves will provide decades of relatively cheap energy for home heating, electricity generation and industry. As a result, coal consumption fell 10 percent in the four years between 2007 and 2011, with natural gas production increasing 15 percent. Since natural gas contains about half the carbon content of the average coal per unit of energy, it emits lower greenhouse gas emissions than coal. Burning natural gas for electricity generation and replacing coal-fired plants results in a reduction of carbon dioxide emissions by nearly a factor of three.[55]

Additionally, natural gas may be the pathway to a reduction in the use of oil in transportation as natural gas–powered trucks, buses, farm equipment begin to power automobiles. Presently, vehicle transportation represents only 0.15 percent of natural gas usage. As Daniel P. Schrag, director of the Center for the Environment at Harvard University, has argued,

> the main impact of shale gas on climate change is neither the reduced emissions from fuel substitution nor the greenhouse gas footprint of natural gas itself, but rather the competition between abundant, low-cost gas and low-carbon technologies, including renewables and carbon capture and storage. . . . Ultimately eliminating the conventional use of coal.[56]

With a stated goal of reducing carbon dioxide emissions by 50 percent by 2050, the electric power–generating capacity of the United States represents the best opportunity to substitute gas-fired power plants for coal-fired plants. Currently, natural gas provides 25 percent of the country's total electric power, but it possesses 40 percent of the total generating capacity. Displacing coal-fired power in the following decades represent the most cost-effective way of reaching the 2050 goal. An additional benefit of using the existing natural gas plants would significantly reduce other pollutants, including sulfur dioxide, nitrous oxide, and mercury, with no appreciable additional cost.[57]

Similar to other fossil fuels, natural gas is a hydrocarbon composed mostly of methane. Methane is one carbon atom joined by four hydrogen atoms (CH_4), which constitutes 97 percent of natural gas. In shale gas production, as little as 3.6 percent to as much as 7.9 percent annually escapes into the atmosphere from venting during hydraulic fracturing. Leaks in pipelines and pressure release valves designed to vent gas over the lifetime of a gas well add to the negative greenhouse gas emissions. As another heat-trapping hydrocarbon, methane is a far more powerful greenhouse gas than carbon dioxide. Unlike carbon dioxide, which will trap heat in the atmosphere for 100 years or more, methane lasts for about 20 years. During that shortened period, however, 1 pound of methane traps as much heat as 72 pounds of carbon dioxide.[58] Even though its heat-trapping power declines over time, leakage caused by subterranean pressures, temperature changes and ground movements caused by drilling nearby reduces the advantages of shale gas over other fossil fuels. When burned, natural gas emits half the carbon dioxide of coal, "but methane leakage eviscerates this advantage because of its heat-trapping power."[59] Recent research challenges the EPA's estimate of methane leakage. New estimates put the rate of methane emissions at 2.3 percent of yearly production or an estimated loss of 13 million metric tons of methane or enough natural gas to heat 10 million U.S. homes.[60]

Conclusion

As the transition of providing light to city streets using whale oil waned, manufactured gas and the infrastructure that promoted its networked use gained favor in cities able to afford the costs. Public lighting spread as the depletion of the world's whale population caused rising costs. In addition, whale oil inefficiencies became observable as lamplighters refueled street lamps individually. The beginning of a centralized lighting infrastructure using manufactured gas improved public safety, made citizens proud of their surroundings and awakened urban neighborhoods to the benefits of nightlife.

Producing manufactured gas by capturing coal gas in a retort, purifying it and distributing it throughout many of the pipelines binding the city together came at a considerable environmental cost. Contaminated land, foul odors and the fear of explosions were among the many negative attributes of manufactured gas. Alternatives most prominently the invention of electric lighting provided a much brighter light than manufactured gas. It proved to be odorless and safer but at a higher cost.

The transition to natural gas for heat and to power a new generation of household appliances and industrial machinery became apparent with the discovery of natural gas reserves with petroleum exploration. Initially, thought of as waste in oil drilling and allowed to burn off, investors and operators capped it and established a regional and eventually a national distribution network. Its expansion ended the era of manufactured gas. Its polluted facilities became superfund sites. Until recent years, coal and oil usage for households and factories dominated the power-generation industry, with natural gas placing behind these energy sources nationally. Hydropower and nuclear energy eclipsed all three sources in selected countries.

The discovery of shale gas and, more important, the development of technology to release it became the biggest energy developments of the early twenty-first century. Compared to conventional natural gas, shale-gas production, transportation, storage and delivery dominates the greenhouse gas footprint for methane's short life of 20 years. Its greenhouse gas footprint is greater than that for conventional natural gas from 22 percent to 43 percent. For the 100-year horizon, methane emissions diminish by the shorter time in the atmosphere. However, in this time frame, shale gas's methane greenhouse gas footprint is 14 percent to 19 percent greater than conventional natural gas's.[61]

The view that methane leaks from shale gas drilling lead to greenhouse gas emissions that are as bad as those produced by burning coal is wrong according to other scientists, including Daniel Schrag. According to him, the shale gas revolution is beneficial to the extent that "the new industry can break the stranglehold that the coal industry has had on the national discussion around climate policy."[62] Shale gas represented only 4 percent of the U.S. supply of natural gas in 2007 and has grown to 48 percent in 2017. U.S. coal consumption fell by more than 15 percent since 2007. The recent bankruptcies of the nation's major coal producers have disrupted its "stranglehold" on discussions about climate policy.

With nearly half the carbon content of coal per unit of energy, natural gas produces half as much carbon dioxide as coal when burned for heat or electricity. Taking into consideration, the newest and most efficient coal plants for generating electricity, the combined-cycle natural gas plant that replaces a coal-fired plant for generating electricity reduced carbon dioxide emissions by nearly a factor of three.[63] Scientific colleagues dispute this conclusion adhering to the view that methane leakage over a 20-year period contributes more to greenhouse gas emissions than does carbon dioxide over 100 years. Professor Schrag argues that it is the 100-year time scale that projections of 20 years are too short in evaluating climate change policies.[64]

Will policy makers and consumers view low-cost natural gas as a bridge to both carbon-free renewables, carbon capture and storage technologies or as an alternative? If viewed as an alternative, then research and development into carbon-free and carbon capture and storage will probably be curtailed. In the years and decades to follow, answers will be forthcoming. By increasing the goal of reducing carbon dioxide by 80 percent would require the complete de-carbonization of the electric power generating capacity. This goal would require the further development of carbon-neutral technologies, including wind, solar, hydro, nuclear and carbon capture and storage.

Notes

1 E.G. Stewart, *Town Gas: Its Manufacture and Distribution*, (London: Her Majesty's Stationery Office, 1958), 43.
2 Ibid.
3 Ibid., 3–4.
4 Ibid., 1–4.
5 Martin J. Hamper, "Manufactured Gas History and Processes," *Environmental Forensic*, 7, no. 1, 55–64.
6 Ibid.
7 Leslie Tomory, *Progressive Enlightenment: The Origins of the Gaslight Industry, 1780–1820* (Cambridge, MA: MIT Press, 2012), 1.
8 Peter Thorshem, "The Paradox of Smokeless Fuels: Gas, Coke and the Environment in Britain, 1813–1949," *Environment and History* 8, no. 4 (November 2002): 382–383.
9 Allen W. Hathaway, *Remediation of Former Manufactured Gas Plants and Other Coal-Tar Sites* (New York: CRC Press, 2012), 35.
10 Joel A. Tarr, "Transforming an Energy System: The Evolution of the Manufactured Gas Industry and the Transition to Natural Gas in the United States (1807–1954)," in

The Governance of Large Technical Systems, ed. Olivier Custard (New York: Routledge, 1999), 20–21.

11 Ibid., 21.

12 Christopher J. Castaneda, *Invisible Fuel: Manufactured and Natural Gas in America, 1800–2000* (New York: Twayne Publishers, 1999), 34–35.

13 Allen W. Hatheway, "World History of Manufactured Gas: A 'world' of Land Redevelopment Possibilities," in *Proceedings of the International Symposium and Exhibition on the Redevelopment of Manufactured Gas Plant Sites*, eds. Stephen James, John Ripp and Dennis Unites (Reading, UK, April 4–6, 2006), 173–179.

14 Joel A. Tarr, "Toxic Legacy: The Environmental Impact of the Manufactured Gas Industry in the United States," *Technology and Culture* 55, no. 1 (January 2014): 110,112–113.

15 Castaneda, *Invisible Fuel*, 58–65.

16 www.hatheway.net/01_history.htm.

17 Tarr, "Toxic Legacy," 125.

18 Thorshem, "The Paradox of Smokeless Fuels," 383.

19 www.hatheway.net/01_history.htm.

20 Michael B. McElroy, *Energy: Perspectives, Problems & Prospects* (New York: Oxford University Press, 2010), 149–151.

21 McElroy, *Energy*, 152.

22 Louis Stotz and Alexander Jamison, *History of the Gas Industry* (New York: Stettiner Bros., 1938), 69–70.

23 Christopher J. Castaneda and Clarance M. Smith, *Gas Pipelines and the Emergence of America's Regulatory State: A History of Panhandle Eastern Corporation, 1928–1993*, (Cambridge, UK: Cambridge University Press, 1996), 19.

24 Christopher James Castaneda, *Regulated Enterprise: Natural Gas Pipelines and Northeastern Markets, 1938–1954* (Columbus: Ohio State University Press, 1993), 15–16.

25 Ibid., 43.

26 Joel A. Tarr, The Next Page: There Will Be Gas: A look at Western Pennsylvania's other natural gas drilling booms (and busts) at www.postgazette.com/news/portfolio/2009/08/02/the-next-page-there-will-be-gas/200908020139.

27 Castaneda, *Invisible Fuel*, 49.

28 The Next Page: There Will Be Gas: A Look at Western Pennsylvania's Other Natural Gas Drilling Booms (and busts).

29 Castaneda, *Regulated Enterprise*, 16.

30 Stewart, *Town Gas*, 44.

31 Castaneda, *Invisible Fuel*, 82–83.

32 Ibid., 46–47.

33 Castaneda, *Regulated Enterprise*, 17.

34 A.C. Monahan, "Natural Gas Big Business," *The Science News-Letter* 63, no. 1 (January 3, 1953): 10.

35 Castaneda, *Invisible Fuel*, 104.

36 Monahan, "Natural Gas Big Business," 11.

37 Mark H. Rose, *Cities of Light and Heat: Domesticating Gas and Electricity in Urban America* (University Park: The Pennsylvania State University Press, 1995), 2, 7–8.

38 This paragraph and those that follow are based on Castaneda, *Invisible Fuel*, 107, 140–146.

39 www.eia.gov/dnav/ng/hist/n9140us2A.htm.

40 Ernest J. Moniz, Henry D. Jacoby and Anthony J.M. Meggs, *The Future of Natural Gas: An Interdisciplinary Study, Executive Summary* (Cambridge, MA: MIT Press, 2011), 5.

41 John H. Herbert, *Clean Cheap Heat: The Development of Residential Markets for Natural Gas in the United States* (New York: Praeger Publishing, 1992), 130–135. Castaneda, *Invisible Fuel*, 187–188.

42 Moniz, et al. *The Future of Natural Gas*, 5.
43 Joseph A. Pratt, Tyler Priest, and Christopher J. Castaneda, *Offshore PIONEERS: Brown & Root and the History of Offshore* Oil *and Gas* (Houston: Gulf Publishing Company, 1997), 214.
44 Tarr, "Transforming an Energy System," 31.
45 "IHS Global Insight, the Economic and Employment Contribution of Shale Gas in the United States," 9 (2011) at www.ihs.com/info/ecc/a/shale-gas-jobs-report.aspx.
46 Tim Flannery, "Fury over Fracking," *The New York Review of Books* LXIII, no. 10 (April 21, 2016): 29–31.
47 Nancy C. Loeb, "The Sand Mines That Ruin Farmland," *The New York Times* (May 23, 2016): A19.
48 Joann Wypijewski, "License to Drill," a review of Eliza Griswald, "Amity and Prosperity: One Family and the Fracturing of America," *The New York Times Book Review* (August 5, 2018): 13.
49 Russell Gold, "Oil and Gas Bubble Up All over," *The Wall Street Journal* (January 3, 2012): A7.
50 Moniz, et al. *The Future of Natural Gas*, 125–128.
51 International Energy Agency, "World Energy Outlook," 2012, 75.
52 Desheng Hu and Shengqing Xu, "Opportunity, Challenges and Policy Choices for China on the Development of Shale Gas," *Energy Policy* 60 (May 22, 2013): 22.
53 Mehmet Melikoglu, "Shale Gas: Analysis of its Role in the Global Energy Market," *Renewable and Sustainable Energy Reviews* 37 (2014): 464.
54 Ibid., 464–466.
55 Daniel P. Schrag, "Is Shale Gas Good for Climate Change," *Daedalus, the Journal of the American Academy of Arts & Sciences* (2012): 72–73.
56 Ibid., 72.
57 Moniz, et al. *The Future of Natural Gas*, 8–9.
58 Anthony R. Ingraffea, "Gangplank to a Warm Future," *The New York Times* (July 29, 2013): A17.
59 Ibid.
60 Ramon Alvarez, Daniel Zavala-Araiza and David R. Lyon, et al., "Assessment of Methane Emissions from the U.S. Oil and Supply Chain," *Science* 361 (July 13, 2018): 186–188.
61 Robert W. Howarth, Renee Santoro, Anthony Ingraffea, "Methane and greenhouse-gas footprint of natural gas from shale formations," *Climate Change* (April 12, 2011), 685.
62 Schrag, "Is Shale Gas Good for Climate Change," 73.
63 Ibid.
64 Ibid., 74–75.

References

Alvarez, Ramon, Daniel Zavala-Araiza, and David R. Lyon, et al. "Assessment of Methane Emissions from the U.S. Oil and Supply Chain." *Science* 361 (July 13, 2018): 186–188. doi: 10.1126/science.aar7204.

Castaneda, Christopher J. *Regulated Enterprise: Natural Gas Pipelines and Northeastern Markets, 1938–1954*. Columbus: Ohio State University Press, 1993.

Castaneda, Christopher J. *Invisible Fuel: Manufactured and Natural Gas in America, 1800–2000*. New York: Twayne Publishers, 1999.

Castaneda, Christopher J., and Clarance M. Smith. *Gas Pipelines and the Emergence of America's Regulatory State: A History of Panhandle Eastern Corporation, 1928–1993*. Cambridge, UK: Cambridge University Press, 1996.

Flannery, Tim. "Fury over Fracking." *The New York Review of Books* LXIII, no. 10 (April 21, 2016): 29–31.

Gold, Russell. "Oil and Gas Bubble Up All Over." *The Wall Street Journal*, January 3, 2012.

Hamper, Martin J. "Manufactured Gas History and Processes." *Environmental Forensics* 7, no. 1 (2007): 55–64.

Hatheway, Allen W. "World History of Manufactured Gas: A 'World' of Land Redevelopment Possibilities." In *Proceedings of the International Symposium and Exhibition on the Redevelopment of Manufactured Gas Plant Sites*, edited by Stephen James, John Ripp and Dennis Unites, 171–181. Reading, 2006.

Hathaway, Allen W. *Remediation of Former Manufactured Gas Plants and Other Coal-Tar Sites*. New York: CRC Press, 2012.

Herbert, John H. *Clean Cheap Heat: The Development of Residential Markets for Natural Gas in the United States*. New York: Praeger Publishing, 1992.

Howarth, Robert W., Renee Santoro, and Anthony Ingraffea. "Methane and Greenhouse-gas Footprint of Natural Gas from Shale Formations." *Climatic Change* 106, no. 4 (April 12, 2011): 679–690.

Hu, Desheng, and Shengqing Xu. "Opportunity, Challenges and Policy Choices for China on the Development of Shale Gas." *Energy Policy* 60 (September 2013): 21–26. doi: 10.1016/j.enpol.2013.04.068.

Ingraffea, Anthony R. "Gangplank to a Warm Future." *The New York Times*, July 29, 2013, A17.

Jamison, Alexander, and Louis Stotz *History of the Gas Industry*. New York: Stettiner Bros., 1938.

Loeb, Nancy C. "The Sand Mines That Ruin Farmland." *The New York Times*, May 23, 2016, A19.

McElroy, Michael B. *Energy: Perspectives, Problems & Prospects*. New York: Oxford University Press, 2010.

Melikoglu, Mehmet. "Shale Gas: Analysis of its Role in the Global Energy Market." *Renewable and Sustainable Energy Reviews* 37 (2014): 460–468. doi: 10.1016/j.rser.2014.05.002.

Monahan, A.C. "Natural Gas Big Business." *The Science News-Letter* 63, no. 1 (January 3, 1953): 10–11. doi: 10.2307/3931638.

Moniz, Ernest J., Henry D. Jacoby, and Anthony J.M. Meggs. *The Future of Natural Gas: An Interdisciplinary Study, Executive Summary*. Cambridge, MA: MIT Press, 2011.

Pratt, Joseph A., Tyler Priest, and Christopher J. Castaneda. *Offshore Pioneers: Brown & Root and the History of Offshore Oil and Gas*. Houston: Gulf Publishing Company, 1997.

Rose, Mark H. *Cities of Light and Heat: Domesticating Gas and Electricity in Urban America*. University Park: The Pennsylvania State University Press, 1995.

Schrag, Daniel P. "Is Shale Gas Good for Climate Change." *Daedalus, the Journal of the American Academy of Arts & Sciences* 141, no. 2 (2012): 72–80. www.jstor.org.ezproxy.neu.edu/stable/23240280.

Stewart, E.G. *Town Gas: Its Manufacture and Distribution*. London: Her Majesty's Stationery Office, 1958.

Tarr, Joel A. "Transforming an Energy System: The Evolution of the Manufactured Gas Industry and the Transition to Natural Gas in the United States (1807–1954)." In *The Governance of Large Technical Systems*, edited by Olivier Custard. New York: Routledge, 1999.

Tarr, Joel A. "There Will Be Gas: A look at Western Pennsylvania's Other Natural Gas Drilling Booms (and Busts)." *Pittsburgh Post-Gazette*, August 2, 2009. www.postgazette.com/news/portfolio/2009/08/02/the-next-page-there-will-be-gas/200908020139.

Tarr, Joel A. "Toxic Legacy: The Environmental Impact of the Manufactured Gas Industry in the United States." *Technology and Culture* 55, no. 1 (January 2014): 107–147.
Thorshem, Peter. "The Paradox of Smokeless Fuels: Gas, Coke and the Environment in Britain, 1813–1949." *Environment and History* 8, no. 4 (November 2002): 381–401. www.jstor.org/stable/20723251.
Tomory, Leslie. *Progressive Enlightenment: The Origins of the Gaslight Industry, 1780–1820.* Cambridge, MA: MIT Press, 2012.
Wypijewski, Joann. "License to Drill." Review of *Amity and Prosperity: One Family and the Fracturing of America*, by Eliza Griswald. *The New York Times Book Review*, August 5, 2018, 13.

6 Nuclear power

Introduction

"On December 2, 1942, man achieved here the first self-sustaining chain reaction and thereby initiated the controlled release of nuclear energy."[1] In the aftermath of World War II in 1945, Lewis L. Strauss, chairman of the newly created Atomic Energy Commission (AEC), expressed optimism about the future of nuclear energy succinctly. He claimed, without qualification, "It is not too much to expect that our children will enjoy electrical energy in their homes too cheap to meter."[2] The civilian use of nuclear power captured the attention of government officials and private utilities. Scientists hypothesized that 1 ton of uranium by fission would release as much energy as 3 million tons of coal.

However, it was the military application of nuclear power that captured the attention of government officials and the public. The detonation of two atomic bombs at Hiroshima and Nagasaki, Japan, by the United States ended the war in the Pacific in 1945. The peaceful uses of atomic energy would be delayed by a geopolitical standoff at the end of the war between two former allies, the Soviet Union and the United States. By 2016, the global nuclear power industry operated 450 reactors, 98 of them in the United States. After a long decline in nuclear power development caused by disasters that are described later, nuclear power generation increased by 1.3% worldwide. With 60 new nuclear power plants under construction globally, only 5 were located in the United States. The resurgence occurred mostly in Asia, with China alone increasing its capacity by 30 percent.[3]

The science and technology of nuclear energy

The science of atomic bombs required the splitting of an atomic nucleus. The net binding energy of protons and their conversion to neutrons determine the stability of the nucleus. As part of the process of rearranging the protons and ejecting neutrons from the nuclei of atoms can cause a chain reaction releasing vast quantities of energy. This process is called fission. Released energy is the source of nuclear bombs and nuclear power.[4] However, knowledge of the potential energy released from the atom came early in scientific investigations. In 1904, the father of nuclear physics, Ernest Rutherford, wrote that the energy of radioactivity came from the atom. His prophetic conclusion signaled the inception of the nuclear age: "If it were ever possible to control at will the rate of disintegration of the radio elements, an enormous amount of energy could be obtained from a small quantity of matter."[5]

Uranium was the only fissionable element from the periodic table that contained the neutrons capable of a chain reaction. Uranium-238 (U-238) is the most plentiful form of this naturally occurring element. It is one of the heaviest elements on the periodic table, almost 19 times heavier than water. Twenty times more plentiful than silver, millions of tons of U-238 exist in Earth's mantle, in open pits and underground. Gold ores, shale, granite and seawater contain fractional amounts. Solvents remove U-238 from mined ore turning it into uranium oxide that undergoes filtering and drying. The result is a substance called yellowcake. Further purification results in the production of uranium dioxide powder. Pressed into pellets and placed into tubes, they become the fuel rods for a nuclear reactor. A nuclear reactor contains 50,925 fuel rods in 193 assembles.[6] When operational, this very complicated process can produce energy without emitting greenhouse gases.

The genesis of the nuclear age, however, delayed peaceful uses of atoms for generating electricity. The age began with the race by Great Britain and the United States to build a nuclear bomb (an atom bomb) during World War II before Nazi Germany achieved success. German émigré scientists supported by the highest-profile scientist, among them, Albert Einstein, convinced President Franklin Roosevelt to create what came to be known as the Manhattan Project in June 1942. The project employed as many as 150,000 scientists and engineers from the United States, Canada and Great Britain and cost more than US$2 billion. Its task, converting nuclear energy into a weapon, proved to test the skills of each country's most talented professionals. They discovered that almost all U-238, the most plentiful form found on the planet, lacked the properties required for a nuclear chain reaction. Breaking it down, however, to capture its isotope U-235 represented the first step in developing a nuclear bomb. Separating one isotope from U-238 would require extraordinary skill, however.

Efforts to achieve this separation revealed that U-238 produced plutonium, a chemical element conducive to a nuclear chain reaction. Suddenly, the separation became secondary because of this discovery. As scientists and engineers discovered, however, making a plutonium nuclear bomb would require a more complicated detonator. As this stage of the discovery and development process, no professionals knew whether an atomic bomb using either uranium or plutonium would achieve its goal of mass destruction.

Without assurances that one element was better than another, the Manhattan Project leaders decided to experiment with both. Government facilities charged with different responsibilities and employing thousands of personnel, spent vast sums of money to become operational. Towns sprang up to house the personnel in separate locations quickly. "For the uranium separation effort, the Manhattan

Project would explore several different methods: gaseous diffusion, electromagnetic separation, and thermal diffusion."[7]

Because of the need for a steady supply of cheap hydroelectric power, Manhattan Project managers chose Oak Ridge, Tennessee for the uranium separation process. It gave them access to the power provided by the many dams built of the Tennessee Valley Authority (TVA) during the Great Depression (1929–1938). The connection between hydropower development and nuclear bomb activities "became powerful symbols of American technological might."[8] To understand the complexity of the separation process at the Oak Ridge laboratories, they used 30 million pounds of silver and built an elaborate and lengthy system of pipes, valves and tubes to withstand the high-pressure separation process. Engineers built similar complex facilities at Hanford, Washington.

There, plutonium separation facilities depended on the workmanship of more than 40,000 workers. They built housing for themselves and others directly involved in technological operations as well as railroads to transport radioactive uranium rods. Hanford's three nuclear reactors and its plutonium separation facilities were spread over a half million acres. Like the Oak Ridge nuclear facilities, Hanford required the hydropower delivered by the many dams built along the Columbia River, including the Grand Coulee Dam built during the Great Depression.

In combination, the nuclear facilities at both sites represented the marriage of hydro- and nuclear power and another symbol of technological prowess. However, Hanford's experiments with plutonium created tons of nuclear waste. In the years that followed, opponents would use this waste stored in tanks belowground as a fundamental reason for opposing both the development of atom bombs and nuclear reactors for delivering emission-free electricity. Decades after the closing of Hanford, the site remained radioactive and off-limits to human activity.[9]

Research and development activities continued at universities, including the University of Chicago, Columbia University and the University of California. The facility at Los Alamos, New Mexico, however, proved noteworthy because on July 16, 1945, at Alamogordo, New Mexico, the first atomic bomb, a plutonium detonated fireball with the power of 20,000 tons of TNT, turned the desert sand into glass. Despite a few years in development, this detonation inaugurated the beginning of the atomic age.

Atoms for peace

Given the ferocity of this new weapon, many scientists hoped that international cooperation would result in its abolition. However, the Soviet Union gained access to nuclear technology and detonated its first atom bomb in 1949. The beginning of the Cold War, with North Atlantic Treaty Organization (NATO) and Warsaw Pact countries facing each other in Europe, dashed the hopes of international cooperation. As a result, enthusiasm became cautious optimism that nuclear power would generate electricity for domestic use.

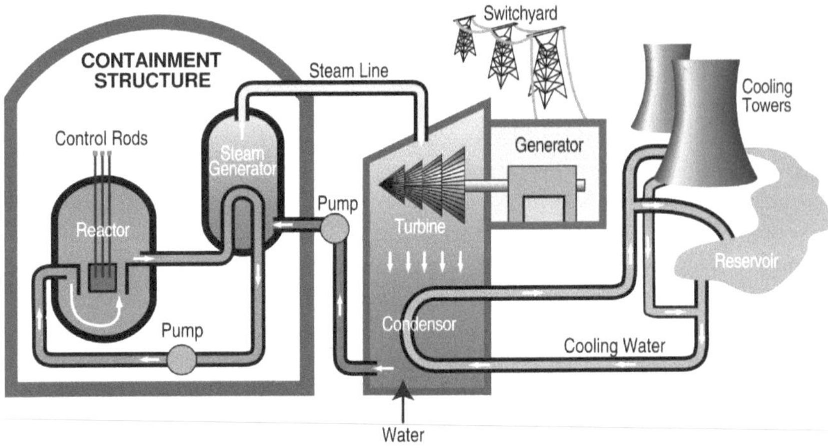

Figure 6.1 A diagram of a nuclear power plant's machinery.

However, David Lilienthal, the former director of the TVA and first chairman of the AEC in the United States, noted that "if the myth that atomic energy is simply a military weapon becomes a fixed thing in our minds, if we accept the error that it can never be anything else, we will never make it anything *but* a weapon."[10] He elegantly voiced the beliefs of many – that atoms for peaceful use, in the form of nuclear power, would produce emissions-free electricity.

However, developing and stockpiling the next generation of nuclear weapons remained a priority of the AEC, which was created by the Atomic Energy Act on January 1, 1947. As a result, plans for the peaceful uses of atomic energy became a less important goal as the AEC pursued upgrades in the country's original reactors and the construction of two new plutonium reactors at the Hanford site. Despite these military developments, the AEC maintained an interest in peaceful uses of atomic energy. Nuclear scientists, many of whom worked on the Manhattan Project, became vociferous advocates of civilian nuclear power. Alvin M. Weinberg, one of the scientists, made his commitment clear by saying that "[a]tomic power can cure as well as kill. It can fertilize and enrich a region as well as devastate it. It can widen man's horizons as well as force him back into the cave."[11]

Dwight D. Eisenhower, president of the United States (1953–1961), spoke before the United Nations General Assembly on December 8, 1953. There, he proposed an international agreement of sharing knowledge about atomic energy among countries capable of producing fissionable materials. Also, nations committed to peaceful uses would receive atomic energy briefings from nations, including the United States and Great Britain. Despite opposition from the Soviet Union, the plan, although never implemented fully, gave the U.S. propaganda advantages over the Soviet Union. The demand for fissionable material might curtail weapons development and take advantage of Great Britain's plans for large-scale nuclear-power generation. In addition, Erratum, a European nuclear cooperative established to lessen an energy-starved continent, represented the implementation of the Atoms for Peace goals. However, the discovery of more productive oil fields in Saudi Arabia, Iraq and Iran lowered the world's price for oil. "When cheap oil became available from the Middle East, support for non-military nuclear power developments on a world scale began to decline."[12]

Another challenge facing civilian control over nuclear power development for domestic use was a growing protest movement. Opposition to stockpiling of atomic weapons grew in the years after Hiroshima and Nagasaki. Atomic bomb testing sites contaminated the surrounding land; send airborne radioactive particles that showed up in food and dairy products. Although opposition remained fragmented, environmental groups, local activists whose NIMBY (Not in My Back Yard) enthusiasts, all raised doubts about safety. However boisterous they became, advances in nuclear technology quickened in the 1960s and 1970s, despite anti-Vietnam War opponents and advocates for the environment.

In the immediate postwar decades (1945–1960), the majority of Americans surveyed supported nuclear power. Nuclear power accidents, however, changed attitudes among the public. A nuclear fire at Windscale, England's Pile Number One, occurred in 1957. Allegedly, nuclear waste buried in the Ural Mountains located in the Soviet Union exploded, killing hundreds of people a year later. In the United States, an experimental nuclear reactor exploded in Idaho Falls in 1961.[13]

Throughout the 1960s, opponents questioned the safety of the reactors' cooling systems. The partial meltdown without the release of radioactive material at Fermi Unit 1 in Detroit, Michigan, in 1961 caught the attention of activists. Administrators closed the unit and dismantled it in 1972. Complaints from citizen groups represented a range of interests. Some focused on reactor safety measures and the dangers of radioactivity released from bomb testing and releases from nuclear reactors. Others, including the Fish and Wildlife Service of the Department of Interior, studied thermal pollution. River and lake water cooled the heated waste of nuclear power plants. Once returned to its source, water temperatures rose by many degrees. The heat killed fish and mollusks that evolved genetically in cooler waters.[14] Despite these protests, applications from private utilities to build nuclear power plants taxed the resources of the AEC.

As environmental historian Martin Melosi has suggested, an ecological worldview, relating all organisms to each other replaced the concept of conservation and preservation. However, as he noted the environmental movement of the 1970s proved to become schizophrenic. On one hand, it applauded the work of scientists and engineers who solved environmental problems. On the other, scientists and technical experts brought the nuclear age to fruition. "For a number of political activists, no technical fix could correct the potential harm of a nuclear accident, and they could settle for nothing less than its abandonment."[15]

In the decades that followed, anti–nuclear power protests became events whenever the AEC approved licenses for private utilities to build nuclear reactors of all kinds, including breeder reactors to produce plutonium. In England, France and Germany, similar protests became commonplace, not so much in the Soviet Union and its satellite countries in Eastern Europe.

The Organization of Petroleum Exporting Countries (OPEC) oil embargo of 1973–1974, caused partially by Western allies supporting Israel in the Yom Kippur War against Egypt and Syria, caused a spike in global oil prices. Renewed interest in civilian nuclear power, clean and renewable peaked enthusiasm for this energy source. Responding to the oil crisis, utilities ordered additional reactors building on the core of nuclear facilities started in the 1950s. Despite the vagaries of licensing power plants throughout the decades that followed, by the late 1970s, the U.S. possessed one-half of the noncommunist world's nuclear-generating capacity. Western Europe followed with one-third, Japan with 11 percent and Canada with close to 5 percent.[16] However, the oil crisis and the global recession that followed and a major spike in inflation, mostly caused by inflated oil prices, depressed consumer demand for electricity.

Reduced demand, the anti–nuclear war and power movement and two potentially catastrophic reactor meltdowns stopped further development in the United States. Three Mile Island (TMI) on the Susquehanna River in eastern Pennsylvania on March 28, 1978, and the nuclear explosion at Chernobyl in the Soviet Union in 1986 riveted the world's attention to the threats of explosions, core meltdowns and spreading radioactivity.

TMI

This power station had two units, TMI-1 and TMI-2. The failure of a pump in unit 2 designed to feed water to a steam generator failed to function properly. A fail-safe computerized system responded by automatically opening a relief value to prevent the reactor from overheating. The panel in the control room, however, failed to recognize this command. The crew staffing the control room thought that the relief value had not been activated and began to drain water from the reactor. Without cooling water, the radioactive core of the reactor overheated, causing a major core meltdown. Despite this catastrophe at unit 2, no explosion occurred limiting the release of radioactive material into the atmosphere. As a result, humans received either limited or no exposure to radiation.

This good news, if one can call it that, did little to comfort a suspicious population. Anti–nuclear power demonstrations took place in 10 large cities. Between 65,000 to 75,000 people rallied at the nation's capitol in protest. A diminished demand for nuclear power and coal-fired plants preceded the meltdown at TMI. From 1979 to the present, nuclear plant cancellations grew, utilities stopped those at the construction stage and the industry remained stuck maintaining the nation's 70 functional nuclear facilities. Because of the events at TMI, evacuation plans to ease citizen fears called for a 5- to 10-mile-radius zone in the case of a meltdown and the release of radiation. The report on TMI issued by the Department of Energy (DOE) stated, "What escaped at Three Mile Island was not only radiation, but, more importantly for the nuclear power industry, public confidence in technology and technocracy."[17]

Chernobyl, Ukraine, Union of Soviet Socialist Republics (USSR)

In April 25–26, 1986, a massive nuclear power accident occurred there. The power station was located about 90 miles from the country's ancient capital, Kiev. Ironically, poorly trained operators began a test of one the power station's four water-cooled reactors by turning off the emergency cooling system. This decision would prevent the reactor from flooding the radioactive core and shutting down the reactor. For reasons that may be explained by changes in the shifts of crew members, they failed to turn on the water-cooling system or activating the automatic power controls. Failing this important function caused power to drop. Observing this drop, the crew tried to restore power by withdrawing radioactive control rods and increasing steam pressure. Computer controls warned of a catastrophe if the crew failed to shut down the reactor. Ignoring the initial warnings, they panicked and lost control of the overheated reactor. Because the unit lacked a containment structure for safety, the reactor exploded tearing off the roof, releasing radioactive debris into the atmosphere. "The explosions ignited graphite in the core and resulted in rampart fires in the reactor building, causing more radiation to be released into the atmosphere. Possibly as much as seven tons of radioactive fuel was blown out of the building."[18]

Aside from the destructive force of the explosion, Ukrainians living in proximity to Chernobyl received elevated radioactive exposure. As many as 180,000 citizens received exposure that required resettlement. Contaminated land and water prohibited habitation. In the short run, the number of deaths included plant employees and firefighters who rushed to Chernobyl to contain the fire's explosive force. In the long term, deaths attributed to radiation exposure remained incomplete despite statistics noting that deaths from cancer in the area serviced by Chernobyl exceeded those found elsewhere. In addition, wind patterns carried radiation to Europe and as far away as Canada and the United States.

The tragedy of Chernobyl empowered anti-nuclear power advocates. As one USSR reporter recalled after visiting the scene, "Chernobyl was a warning for the future. It was not just a banal disaster, it was a message that nuclear power was not safe."[19] Despite improvements in design with multiple computer-coded mechanisms to prevent another Chernobyl, the matter of safety continued to plague the industry. The industry felt the impact of the meltdowns at TMI and Chernobyl. A breakdown in the safety system of one nuclear power plant caused a chain reaction throughout the industry. To assuage the public's fears, the U.S. Congress updated and renewed the Price–Anderson Act and capped liability at US$7 billion up from US$700 million. This legislation reflected the strong anti–nuclear power sentiment in the U.S. Congress and placed additional financial liability on the nuclear power industry.

Fast-forward to the second decade of the twenty-first century and the nuclear power industry has experienced some growth. Now, 61 commercial nuclear power plants, with 99 nuclear reactors provide electricity in 30 states. The most recently built and operational one began service in October 2016 at Watts Bar, Tennessee. As the first nuclear power plant to become operational in this century, it is projected to produce 700 billion kilowatt-hours of renewable energy during its lifetime. By a recent count, all reactors now in operation generated 805 billion kilowatt-hours of electricity, or 19.5 percent of the nation's electric generation.[20]

Fukushima

The impact of Tohoku Region Pacific Coast Earthquake and Tsunami that occurred on March 11, 2011, caused a meltdown and explosion at the Fukushima Daiichi No. 1 Nuclear Power Station on the Pacific coast. The 16,000 deaths and 3,500 people unaccounted for made it a major nuclear power plant disaster. Its impact on Europe's nuclear footprint and its development plans for 2020 stalled in the wake of Japan's catastrophe. Unlike Chernobyl, however, an earthquake and tsunami caused death and destruction, not exposure to radiation.

Pause in Europe led to revision and reversal. Belgium, Germany and Italy decided to phase out their nuclear capacity, culminating in a complete shutdown in the second and third decade of the century. By 2025, nuclear energy will provide 50 percent rather than the planned 75 percent. France followed its European partners by scrapping plans, shutting down plants and allowing the construction of only one nuclear power plant.[21] After the Fukushima disaster, the European Union imposed a series of protocols called *stress tests* to ensure power plant safety. These tests were described as "a targeted reassessment of the safety margins of nuclear power plants in the light of the events which occurred at Fukushima."[22] The future design of power plants, the reduction of accidents and safe operations became the standards for nuclear safety. Although none of the operational 145 reactors in 64 power plants were ordered to shut down, all would require expensive retrofitting to meet the new stress test standards.

The EU Energy Roadmap 2050 promulgated an aggressive goal of reducing greenhouse gases by 80 and 95 percent below 1990 levels. Although nuclear power will play an important role in the EU formulation, it will represent no more than 25 percent of Europe's total energy supply by 2050.[23] EU partners operate 143 of the world's 450 nuclear power reactors. However, the European Union's restrictive new regulations call for the construction of only 20 new nuclear reactors while the developing world, as we will see later in this chapter, forges ahead.

Civilian uses around the world

In Europe, during an 18-year period, from 1973 to 1990, civilian nuclear power expanded at the rate of 18 reactors a year, each generating 16,000 megawatt-hours annually. TMI, Chernobyl and falling prices for fossil fuels deterred future investments. In the 1990s, Europeans built only four reactors each year, generating 3,000 megawatt-hours of electricity. In generating about 30 percent of its electrical needs from nuclear power, Europe produced as much as twice the world average.

Only Denmark and Norway export more energy than they consume, while the rest of the continent imports 53 percent of its energy needs, with coal, oil and natural gas representing almost 70 percent of the continent's needs. EU energy and climate goals for 2020 have targets of 20–20–20, meaning that EU countries will reduce greenhouse gases by 20 percent below 1990 levels, that 20 percent will come from renewable sources and another 20 percent from improved efficiency.[24] Sixteen out of 27 EU countries use nuclear energy for power. Across the continent in 2012, 185 nuclear power plants generated 162 Gigawatt-hours. Currently, France leads with 77.7 percent of nuclear power, followed by Belgium and the Slovakian Republic with 54.0 percent and Ukraine with 47.2 percent.

United Kingdom

According to government sources in the United Kingdom, meeting its target of reducing greenhouse gases by 80% by 2050 required the expanded use of nuclear power. Opposition leaders argued that carbon capture technologies along with expanded wind and solar installations would help in meeting the 2050 target reductions. However, the UK possesses a complicated regulatory apparatus governing nuclear power. Divided powers situated in many agencies thwart nimble, straightforward decision-making. Additionally, Scotland, an independent legal entity in the United Kingdom, controls nuclear power planning and installations. Scotland's parliament possesses the power, recognized by Britain's parliament to block any expansion of nuclear reactors within its territorial boundaries. Understandably, a stalemate best explains Britain's policy on nuclear expansion.

Finland

In Finland, another example of a member of the European Union committed to expanding its nuclear profile built two new reactors bringing the country's total to six. Its government argued that in addition to reducing its carbon footprint, building new reactors would support Finland's industry. As expected, the events at Fukushima dampened support for nuclear power among the country's citizens. Following the accident, only 38% of the public supported the construction of new reactors. Another argument against the cost of building additional nuclear capacity stated that coal and natural gas purchased from Russia would be lower than the costs of nuclear power. An International Energy Agency (IEA) study suggested that rising fossil fuel prices would be accompanied by an expansion of wind and solar power development.[25]

France

France is the third EU country committed to increasing its nuclear power capacity. France maintains 58 nuclear reactors with a capacity of 63 Gigawatt-hours. They represent 78 percent of the country's total electrical output. Unlike many countries in the European Union, France's nuclear power industry is state-owned, and authority is centralized in one agency, the Commissariat a l'Energie Atomique et aux Energies Alternatives. As noted in its title, it controls all military and civilian applications of nuclear power.

Although public opinion turned negative after Fukushima, France's centralized decision-making authority muted most negative public opinion. Once the news of Fukushima receded in the public's consciousness, two-thirds of the population supported nuclear power. With centralized control, the government worked successfully to point out the benefits of nuclear energy. It emphasized national security, environmental progress, waste management and a history of social acceptance of nuclear power. Along with agency controls, presidential power over nuclear decisions is somewhat unique when compared to other EU members. As one expert noted, "France is probably the only country in the world where the President can say 'we will have a reactor here.' "[26]

China

While the Fukushima disaster caused many countries in the West including Germany, Belgium, Italy and Switzerland to phase out their nuclear power plants, China, and many of Asia's countries, paused and then forged ahead. China's growth in nuclear power development outstripped the rest of the world. Its first nuclear power station opened in 1991. By 2000, its power capacity reached 2.1 gigawatt-hours. By 2020, China plans a nuclear power supply of 40 gigawatt-hours with an additional 18 gigawatt-hours in the construction phase.[27] Having implemented the largest program in nuclear power construction, Chinese citizens experienced doubts similar to those held by Europeans about safety. Although survey findings were mixed, data noted that Chinese citizens living closer to the Fukushima site believed that they were at greater risk from nuclear accidents. Risk perception was greater among women, citizens over the age of 35, and those with college degrees.[28]

Despite China's initiative to make the transition to renewable energy and away from fossil fuels, the current state of affairs is sobering. About 70 percent of the country's primary energy consumption is by burning coal. Although efforts to reduce coal consumption will continue in the coming decades, China's dependence on fossil fuels will remain its primary source of power generation. Its research on clean-coal technology suggests that it knows that a carbon footprint will remain a significant part of its energy mix in future decades. Without sequestering carbon and developing clean-coal technologies, the death toll from burning coal will continue to affect China's population. On average about half a million people die prematurely from the country's polluted air. The country's goals to reduce fossil fuel emissions are driven by its clean-air initiatives.

"Nuclear energy is not an option, but a necessity"

The title of this section, a quote from Sin Qin, China's president of its National Nuclear Corporation, states the obvious for China. There, constructing 40 percent of the world's nuclear reactors (26 out of 65) represents one important phase in the country's quest for energy security and meeting its planned reductions in carbon dioxide, sulfur dioxide, and nitric oxide. Without China's current nuclear power capacity of 10.8 gigawatt-hours, it would be emitting an extra 67 million tons of carbon dioxide. Also, nuclear power possesses an efficiency factor of from 70 to 90 percent, while the factor for coal-fired plants ranges from 50 to 60 percent, hydropower from 30 to 40, and wind or solar energy from 20 to 30.[29]

China's economic growth during the last decades, reaching 10 percent and more on an annual basis contributed greatly to the emergence of the country's middle class. Globally, growing middle classes place high demands on national services and expectations for improved living conditions that include modern housing with updated appliances. Infrastructure in the form of modern roadways for newly purchased automobiles, an international symbol of middle-class life, high-speed rail systems, airports and the like signify economic growth and upward mobility. All place demands on China's electrical capacity. As noted, burning more coal to generate electricity for the country's growing middle class contaminates the air and creates a public health crisis. The government is committed to reducing fossil fuel consumption and replacing it with renewable sources, primarily nuclear reactors.

The expansion comes with a host of challenges. As is the case almost everywhere, China's population is not geographically distributed across the country. With the majority living near or in coastal regions, the matter of building electrical grids to meet growing demands becomes problematic. Technical expertise, skilled labor, construction, nuclear fuel management and nuclear security to prevent acts of terror, all became simultaneous priorities. To promote the highest standards, China's nuclear industry is part of the global nuclear industry. Its 1996 *Regulation on the Nuclear Industry* adopted much of the international code for regulations and safety. Its reactors are among the world's most advanced, and its thousands of engineers, technicians and managers have received training in the United Kingdom, France, Germany, Canada and the United States.[30]

India

Much like China's, India's construction programs for nuclear power plants are large when compared to other parts of the world. For both countries, energy needs dwarf current operational nuclear reactors and those in either the construction or planning stages. With a centralized government and China's economic growth, the capital needed for a rapid expansion seems to be available. India's democratic government, its rapid population growth and its lack of surplus capital to fund nuclear power expansion may retard its needed demand for electricity. Regardless of the different developments for China and India, Japanese scientist Kunihiko Uematsu argued that "the future development of this energy source may no longer be spearheaded by the traditional developed countries of Europe and North America."[31]

Although significant future developments of nuclear energy will probably take place in Asia, India and China have the longest road ahead in the transition from fossil fuels to renewable sources. Unlike France (75 percent), South Korea (40 percent) and Japan (35 percent), nuclear power makes a significant contribution to their power supplies. In India, as in China, coal will remain the only fossil fuel in abundance even though its nonrenewable status means that someday in the distant future both countries will deplete their supplies.[32]

Like China's, India's abundance of cheap, low-grade coal with a high fly-ash content of between 55 and 60 percent is a less efficient fuel source. Because most of the coal comes from the country's east, long-haul trains burn diesel to transport it over long distances. Burning low-efficiency coal with an average thermal capacity of from 35 to 40 percent exacerbates India's poor air quality and has an adverse impact on its people, environment and ecology. Cooking with low-grade coal adds to its poor air quality. Like China, research to discover ways to capture the carbon content of its coal will remain a high priority.

India imports about 70 percent of its oil, placing a drain on its currency. Burning coal and oil supplies 64 percent of the country's thermal power. For India, burning more coal became a national priority. Going forward to 2020, it would burn 1.5 billion tons a year, adding greatly to its polluted air and to global climate change. Currently, it is the world's third-largest emitter behind China and the United States yet fails to meet its energy needs. Nearly a quarter of its 1-billion-plus population survive with no electricity while others receive it irregularly. The availability of domestic natural gas is negligible, although future supplies from the Russian Federation remain a promising addition to India's energy needs. Hydropower, as controversial in India as in other countries because of the displacement of indigenous people, provides about 25 percent of its electrical power needs.[33]

For nuclear power in India, currently supplying about 4 percent of its thermal power, the road ahead will remain challenging, despite goals affirming the role of nuclear power in the country's future. The world average for this power source remains about 16 percent, making efforts to catch up more challenging. Among the most articulate statements supporting investments in nuclear power is the following, "there is a consensus, across the major political parties, that given India's existing and future energy needs, nuclear power provides a potentially attractive source."[34]

The goal of providing 20 gigawatt-hours of electricity by 2020 is a daunting one considering that population growth and economic development will not change the 4 percent contribution to energy security. The more distant goal of installed nuclear capacity of 275 gigawatt-hours by 2050 will provide 25 percent of the country's energy needs, if it delivers on its promise of a robust nuclear development program. Given India's modest reserves of uranium, to achieve the 2050 goal will require the deployment of fast nuclear reactor technology. What does this mean? Currently, India deploys pressurized heavy water reactors (PHWRs) that are fueled by natural uranium, the element in short supply. The next step will be achieved by deploying plutonium breeder reactors. The PHWRs will produce plutonium to fuel the new reactors. Finally, and probably not operational until 2050, will be the deployment of thorium reactors. Since India possesses a large supply of thorium, nuclear fuel self-sufficiency promises to become a reality by mid-century.[35]

India receives the equivalent of 5,000 trillion kilowatt-hours of solar energy every year, more than the total energy the country consumes.[36] The steep drop in the price of renewable solar energy made tapping into this enormous potential energy source a viable alternative to fossil fuels. A commitment to the 2016 Paris Agreement to reduce greenhouse gases reinforced the change in direction. Now India's goal strives to achieve 40 percent of its electricity from renewable sources by 2030. Utilities stopped building coal-fired plants while government actions prohibited renewing licenses for aging ones. Solar power now costs less than coal without factoring in the benefits of cleaner air and a reduction in emissions. Although coal will continue to dominate the thermal power market, the addition of third-generation nuclear reactors and a steady increase in generating electricity from solar power offers a way forward without continuing dependence on fossil coal.[37]

Republic of Korea

Coal and nuclear power provide 70 percent of the country's electricity, with nuclear providing one-third of the total. With most of its coal imported, the density of nuclear power, given the country's small size, is unique among nations with a nuclear power footprint. With 25 nuclear plants in 2016, it was the fifth-largest producer in the world behind China. Korea's geographical location and its proximity to Japan heightened its reaction to the Fukushima nuclear meltdown. In addition, two earthquakes measuring 4.5 to 5.8 magnitude in the southeastern section of the country exacerbated public reaction. Knowingly, the government built nuclear reactors adjacent to a geological fault line and capable of withstanding a 7.0-magnitude earthquake. Repeated earthquakes of a smaller magnitude, however, could damage Korea's prominent southeast nuclear footprint. There, it located 70 percent of its nuclear reactors far away from its enemy The People's Democratic Republic of Korea, where a truce never technically ended the Korean War (1950–1953).

Despite the impact of the Fukushima disaster on public attitudes, general support for nuclear power remained positive at between 87 and 90 percent. Local acceptance, meaning proximity to a reactor, wavered considerably, however. Generally, never higher than 28 percent, local support dropped to 20 percent in the wake of Fukushima. More important, it affected government policy. The government scrapped its plan to build an additional eight reactors in the next 15 years in addition to the four units under construction. In June 2017, the Republic of Korea's president, Moon Jae-in, made this announcement and added that it will not relicense existing plants. Old coal power plants will be shut down, and plans for new ones canceled.

In their place, plans to increase natural gas, hydropower and solar are viewed as alternative to coal and nuclear. However, rapid replacement and transition to fewer greenhouse sources of power face some problems. The issue of costs during such a rapid transition makes it problematic, however. Imposing an environmental tax on coal and nuclear plants remains a goal of President Moon's administration. Reducing or eliminating import tariffs on natural gas is another way to ease the transition but with costs.

Presently, natural gas provides the Republic of Korea with 18 percent of its energy. Reaching 27 percent by 2030 would represent a significant increase. Currently, solar power, now at 5 percent, would jump to 20 percent by 2030. Its continued economic growth requires investments in electrical generation. If the current plans fail to deliver the demand for electric power, producers and consumers suffer the consequences. With no guarantee that natural gas and solar prices will remain low, replacing coal and nuclear too quickly may damage the country economy.[38]

Japan

Much like France, Japan possesses few energy resources. With few domestic sources of coal, oil, natural gas or uranium, Japan must import more than 80 percent of its primary energy resources. As noted on the website of The Federation of Electric Companies in Japan, "[e]nergy is the 'life blood' of any economy, but for Japan, this truism takes on added importance. Just a few simple facts should help to make it clear that Japan is poor in natural resources, specifically sources of energy, which are vital to a healthy, modern economy."[39]

Recognition of a fuel crisis became apparent by the end of the Great War (1914–1918) as the country began a rapid industrial and manufacturing expansion. Additionally, Japan, unlike other energy-poor countries, is earthquake country. As a seismically active part of the "chain of fire," Eurasian and Pacific tectonic plates grind against each other, causing earthquakes and volcanic eruptions. Both advocates of nuclear power and its critics in Japan and around the world paused and noted with alarm the devastation, loss of life and power caused by the earthquake and tsunami that destroyed the Fukushima nuclear facility in March 2011.

As other nations watch aghast from afar, the Japanese experienced the devastation firsthand. Yet, the country's commitment to nuclear power that commenced within a decade after the end of World War II (1939–1945) remains a priority. In 1955, Japan's Atomic Energy Act became law, committing the country to the peaceful used of atomic energy. Much like a similar Atomic Energy Act passed in the United States in 1954, they committed it to peaceful uses. The extended Cold War with the former Soviet Union that lasted until 1991 and the nuclear weapons competition between the two muted the pledge for the peaceful uses of atomic energy. By 1965, Japan owned an operational nuclear power plant. The Fukushima power plant began commercial service in 1970. By longevity standards, Fukushima was an aging nuclear facility when the tsunami hit.

Since 1945, nuclear energy has played an increasingly important role in Japan's economic development. Self-sufficiency in energy without nuclear power is among the lowest in the industrialized world at 4 percent. With its commitment to nuclear since 1954, the ratio increased to 20 percent of Japan's power. Before the Japanese felt the impact of the Fukushima meltdown, 54 nuclear reactors provided power. By January 2017, only four remained operational, and the government plans to decommission 11 more. Six of these were part of the Fukushima nuclear complex. Some planned restarts in 2017 met with considerable opposition from residents near the reactors, with as many as 60 to 70 percent opposing such plans.

With many reactors reaching the end of their 40-year time limit, decisions about the future of nuclear energy remain in flux. Natural gas imports rose to more than 15 percent and oil imports fell to a bit less than 10 percent. Public interest rose in the nonpolluting renewable sources, such as wind and solar. However, they presently account for only 1 percent of Japan's primary energy supply. To compensate for the variability of solar and wind, fuel cell technology using hydrogen and oxygen to generate power dominates thinking about the next generation of fuel cell–powered automobiles. As an innovator in automobile technology, Japan plans to introduce 50 million fuel cell–powered automobiles by 2020 in the world market. Without the unqualified support for nuclear energy in the wake of the Fukushima disaster, this newer technology may become Japan's contribution to the transition from nuclear energy to renewable fuel cell technology.[40]

The Russian Federation

Within the political boundaries of the federation, Russia's energy reserves place it first in natural gas, second in coal and eighth in oil reserves. With so much of this vast country's unexplored land, its fossil fuel reserves may be greater than we know now. Russia is the world's largest energy exporter, with natural gas and oil representing more than 36 percent of the country's budget. Changes in the world market for fossil fuels affect Russia, with a gross national product equal to New York State, more directly than many of the world's other economies.

It also produces nuclear energy and expects to export its technological knowhow, its fast neutron reactors and its research about dealing with nuclear waste and storage. More than 20 such reactors are currently planned for export, with other countries paying US$133 billion for them. The initiative to lead the world's premier nuclear construction program became a Cold War goal with prominent military applications after its successful performance as an ally during World War II. By the 1960s, generating nuclear power for commercial and industrial purposes became a priority as well. Then came the Chernobyl meltdown in 1986 followed by the collapse of the Soviet Union in 1991.

Despite these setbacks, the former Soviet Union possessed the world's largest supply of uranium, much of it produced by former Eastern European satellite countries. The collapse of the Soviet Union in 1991 eliminated these sources of supply. On its own, Russia has an estimated 90-year supply for its own use and exports uranium to countries facing shortages in its domestic production. Coupled with a policy initiated in the 1990s of recycling spent fuel from breeder reactors, Russia produces each year at least 20 tons of potentially weapons-grade plutonium for its defense program. Transferring plutonium to its civil nuclear power industry engages Russia and the United States in complicated international negotiations.

Despite unanswered questions about nuclear safety at home caused by the Chernobyl meltdown and its aftermath, government policy aggressively promotes nuclear energy for civilian purposes at home and abroad. Its multibillion-dollar plan to make Russia a global supplier of nuclear power began when state-owned nuclear company Rosatom took the lead in developing and selling fast-breeder and small modular nuclear reactors. Its 2030 goals call for building 40 new reactors for the federation and using its technological knowledge and manufacturing capacity to fill approximately 80 orders from other countries. In the wake of Fukushima, Germany abandoned its nuclear program, Japan seems stunned by the disaster but moved forward and the program in the United States looks moribund. Russians regard these events as an opportunity to seize a position of leadership in exporting nuclear reactors.[41]

Russia's reach extends into the heartlands of the American West. It owns uranium mines in Wyoming and supplies one-half of the nuclear fuel used by U.S. reactors. Rosatom identifies American and British markets as potential consumers of its reactors. Its primary market, however, will remain the developing world, with an offer that exceeds that from other developed countries. Called one-stop shopping, Russia intends to build reactors for its customers, sell them nuclear fuel and accept spent fuel from its reactors. This last stage eliminates the need for developing countries to build geologic waste repositories.[42]

Many of its customers find Russia's leadership in the development of fast-breeder reactors appealing. Boiling water and pressurized-water reactors consumed enriched uranium and produced waste that remained radioactive for thousands of years. This new generation of reactors recycles spent fuel in the following way. By consuming fissionable enriched uranium for energy, these new reactors create neutrons that collide with residual uranium. This low-grade uranium is transformed into "breeding" plutonium that the reactor burns for energy. In the process, the spent plutonium generates its own radioactive waste. Complicated as it may seem, the important point to remember is that this new generation of reactors can produce 10 to 100 times more energy than earlier-generation nuclear reactors.[43]

As with any new technology, there may be known drawbacks and possible unintended consequences. Plutonium production gives a country access weapon's grade nuclear material. Proliferation of this material carries with it the threat of a conflict using nuclear weapons. Additionally, the cores of fast breeders use liquid-sodium coolant rather than water to prevent overheating. In case of trouble, boiling or pressurized water from reactors can be exposed to air. Sodium ignites when exposed to the air, causing fires and explosions that could send radioactive material into the atmosphere.

Safety measures become a priority in the operation of all nuclear reactors, including the new generation of breeder reactors. Since Chernobyl in March 1986, some researchers suggested that the Soviet Union/Russia undervalued safety. The disaster resonated across the country and the world. Some scholars believed that

Figure 6.2 Alternative energy concept with wind turbines, solar panels and nuclear energy power plants.

along with the Soviet Union's war in Afghanistan (1979–1989), the Chernobyl meltdown created economic conditions that led to the demise of the Soviet Union in 1991.[44]

Conclusion

Proponents of nuclear energy believe that it must be in the mix of energy sources going forward. Most of the world's 6 billion poor people need electricity to begin the process of rising economically. So, over the next 50 years, its advocates will promote nuclear energy as clean and safe is clear. Compared to wind and solar, discussed in the chapters to follow, nuclear energy does not depend on sunlight or wind to generate electricity. In addition, new generations of small reactors currently at the experimental stage will replace those presently in use. The advanced technology of these reactors will ensure efficiency and safety. A carbon-free energy source has much to celebrate when compared to the effects of greenhouse gas emissions from coal, oil and natural gas on the planet's rising temperatures.

The opponents of nuclear energy argue that the construction, licensing and operation of a nuclear power plant take seven or more years. During this period, carbon emissions from fossil fuels will continue to raise the planet's temperature. For nuclear, mining uranium, building the power plants, providing water to cool the reactors and disposing of radioactive wastes add carbon emissions to the atmosphere. Nuclear power plants have a larger footprint than wind and solar farms, each of which takes about two years to build and make operational.

Despite the number of solar panels or wind turbines installed to generate electricity, their spacing, not their footprint, makes them more feasible than nuclear plants. Finally, producing weapons-grade plutonium makes the world a more dangerous place. Both India and Pakistan developed nuclear weapons before making the transition to peaceful uses. North Korea has developed nuclear weapons without peaceful uses. Before the accord between Iran, Russia, China and the United States in 2016 restricting Iran's capacity to develop nuclear weapons, it was on the path to becoming another nuclear power.

China, India and others recognize the need to deeply decarbonizes the global energy system in order to meet the targets set by the Paris Agreement to limit temperatures increases to "well below 2 °C (1.8 °F) above preindustrial levels."[45] China is building 26 commercial nuclear reactors, including two of the largest ever assembled. In the United States, one of the world's largest emitters of greenhouse gases, it appears that it will lose much of its nuclear power capacity over the next few decades and a source of reliable carbon-free energy. Public opposition to nuclear power and economic pressure from low natural gas prices have led to the closure of some nuclear power plants and the abandonment of partially constructed ones. Those existing large light-water reactors will require costly refurbishment and upgrades in the billions of dollars and advanced fission designs will unlikely be available until mid-century. By that time, the world will probably have missed the 2° Celsius target, and a crash program will be needed to curtail future uncontrollable global warming.[46]

Notes

1 Text from a plaque located on the squash court at Alonzo Stagg Field on the campus of the University of Chicago where the nuclear physicist, Enrico Fermi, performed the first nuclear reaction.
2 *The New York Times*, September 17, 1954.
3 James Conca, "The Year in Nuclear," https://www.forbes.com/sites/jamesconca/2017/01/03/2016-the-year-in-nuclear/#1dfd9e741c89 2/66.
4 Michael B. McElroy, *Energy: Perspectives, Problems, & Prospects* (New York: Oxford University Press, 2010), 193.
5 Sir John Cockcroft, *The Development and Future of Nuclear Energy* (Oxford: The Clarendon Press, 1950), 4.
6 Ibid., 196.
7 Audra J. Wolfe, *Competing with the Soviets: Science, Technology, and the State in Cold War America* (Baltimore, MD: The Johns Hopkins University Press, 2013), 11.
8 Ibid.
9 Ibid., 13.
10 David Lilienthal, *Change, Hope and the Bomb* (Princeton: Princeton University Press, 1963), 77.
11 Quoted in Nuclear Regulatory Commission, *A Short History of Nuclear Regulation, 1946–1999*, 3 (January 29, 2001), www.nrc.gov/SECY/sm/shorthis.htm.
12 Martin V. Melosi, *Atomic Age America* (Upper Saddle River, NJ: Pearson Education, Inc., 2013), 161.
13 Ibid., 226.
14 Ibid., 230.
15 Ibid., 231.

16 Ibid., 245.
17 Ibid., 243.
18 Ibid., 263–264.
19 Ibid., 267.
20 U.S. Energy Information Administration (EIA) at www.eia.gov/faq.php?id=207%t=3.
21 Ibid., 064001–2.
22 Ibid., 064001–4.
23 Ibid., 064001–5.
24 Luc H. Geraets and Ir. Yves A. Crommelynck, *Journal of Pressure Vessel Technology* 134 (December 2012): 064001–1.
25 International Energy Agency. 2008. "World Energy Outlook." At http:/www.iea.org/textbase/nppdf/free/2008/weo2008.pdf.
26 Keith Baker and Gerry Stoker, "Metagovernance and Nuclear Power in Europe," *Journal of European Public Policy* 19, no. 7 (2012): 15.
27 Jin Yang, Donghui Zhang, Mi Xu, and Jinying Li, "China's Nuclear Power Goals Surge Ahead," *Science* 340, no. 6129 (April 12, 2013): 142.
28 Lei Huang, Yuting Han, James K. Hammitt and Yang Lui, "Effects of the Fukushima Nuclear Accident on the Risk Perception of Residents near a Power Plant in China," *Proceedings of the National Academy of Sciences of the United States of America*, 110, no. 49 (December 3, 2013): 19747.
29 Y.C. Xu, "The Struggle for Safe Nuclear Expansion in China," *Energy Policy* (June 16, 2014): 1.
30 Ibid., 27.
31 David Bodansky, *Nuclear Energy*, 2nd ed. (2004, New York: Springer Verlag), 604.
32 "Future of Nuclear Power in India," *Current Science* 102, no. 5 (March 10, 2012): 659.
33 U.C. Mishra, "Environmental Impact of Coal Industry and Thermal Power Plants in India," *Journal of Environmental Radioactivity* 72 (2004): 35.
34 M.R. Srinivasan, R.B. Grover and S.A. Bhardwaj, "Nuclear Power in India: Winds of Change," *Economic and Political Weekly* (December 3, 2005): 5183.
35 M.V. Ramana, "Nuclear Power in India: Failed Past, Dubious Future," *The Nonproliferation Policy Education Center* (May 10, 2006): 2.
36 "India Should Exploit Renewable Energy," *Nature* 145 (January 12, 2012): 145.
37 Anand, G. "India, once a Coal Goliath, Is Fast Turning Green," *The New York Times*, https://nyti.ms/2sx0qp9 and www.nytimes.com/2017/06/02/world/asia/india-coal-green-energy-climate.html.
38 "South Korea Scraps Plants, Signals Shift from Nuclear Energy," *The New York Times* (June 19, 2017), 1, https://nyti.ms/2sFUc4R.
39 www.japannuclear.com/getting started/gettingstarted.html.
40 Masuro Ogawa and Tetsuo Nishihara, "Present Status of Energy in Japan and HTTR Project," *Nuclear Engineering and Design* 233 (2004): 1.
41 Eve Conant, "Russia's New Empire," *Scientific American* 309, no. 4 (October 2013): 88–89.
42 Ibid., 89.
43 Ibid.
44 Yelizaveta Mikhailovna Sharonova, "Nuclear Energy Status in Russia: Historical and Contemporary Perspectives," *Journal of Humanities and Social Science* 21, no. 9 (September 2016): 39–41.
45 T.F. Stocker eds., et al., *Climate Change 2013: The Physical Science Basis. Contribution of Working Group1 to the Fifth Assessment Report of the Intergovernmental Panel on Climate Change* (Cambridge, UK: Cambridge University Press).
46 M. Granger Morgan, Ahmed Abdulla, Michael J. Ford, and Michael Rath, "US Nuclear Power: The Vanishing Low-carbon Wedge," *Proceedings of the National Academy of Sciences* 115, no. 28 (July 10, 2018): 7184–7189.

References

Anand, G. "India, Once a Coal Goliath, Is Fast Turning Green." *The New York Times*, June 2, 2017. https://nyti.ms/2sx0qp9 and www.nytimes.com/2017/06/02/world/asia/india-coal-green-energy-climate.html.

Baker, Keith, and Gerry Stoker. "Metagovernance and Nuclear Power in Europe." *Journal of European Public Policy* 19, no. 7 (2012): 1026–1051. doi: 10.1080/13501763.2011.652900.

Bodansky, David. *Nuclear Energy: Principles, Practices, Prospects*. New York: Springer-Verlag, 2004.

Cockcroft, Sir John. *The Development and Future of Nuclear Energy*. Oxford: The Clarendon Press, 1950.

Conant, Eve. "Russia's New Empire." *Scientific American* 309, no. 4 (October 2013): 88–93. doi: 10.1038/scientificamerican1013–88.

Conca, James. "The Year in Nuclear." *Forbes*, January 3, 2017. https:www.forbes.com/sites/jamesconca/2017/01/03/2016-the-year-in-nuclear/#1dfd9e741c89 2/66.

Geraets, Luc H., and Ir. Yves A. Crommelynck. "Nuclear Power in Europe 2012." *Journal of Pressure Vessel Technology* 134, no. 6 (December 2012). doi: 10.1115/1.4007979.

Huang, Lei, Yuting Han, James K. Hammitt, and Yang Lui. "Effects of the Fukushima Nuclear Accident on the Risk Perception of Residents Near a Power Plant in China." *Proceedings of the National Academy of Sciences of the United States of America* 110, no. 49 (December 3, 2013): 19742–19727. doi: 10.1073/pnas.1313825110.

Lilienthal, David. *Change, Hope and the Bomb*. Princeton: Princeton University Press, 1963.

McElroy, Michael B. *Energy: Perspectives, Problems, & Prospects*. New York: Oxford University Press, 2010.

Melosi, Martin V. *Atomic Age America*. Upper Saddle River, NJ: Pearson Education, Inc., 2013.

Mishra, U.C. "Environmental Impact of Coal Industry and Thermal Power Plants in India." *Journal of Environmental Radioactivity* 72 (2004): 35–40. doi: 10.1016/S0265-931X(03)00183-8.

Morgan, M. Granger, Ahmed Abdulla, Michael J. Ford, and Michael Rath. "US Nuclear Power: The Vanishing Low-Carbon Wedge." *Proceedings of the National Academy of Sciences* 115, no. 28 (2018): 7184–7189. doi: 10.1073/pnas.1804655115.

Nuclear Regulatory Commission. "A Short History of Nuclear Regulation, 1946–1999." Accessed November 4, 2018. www.nrc.gov/SECY/sm/shorthis.htm.

Ogawa, Masuro, and Tetsuo Nishihara. "Present Status of Energy in Japan and HTTR Project." *Nuclear Engineering and Design* 233 (2004): 5–10. doi: 10.1016/j.nucengdes.2004.07.018.

Ramana, M.V. "Nuclear Power in India: Failed Past, Dubious Future." *The Nonproliferation Policy Education Center* (May 10, 2006): 2.

Sharonova, Yelizaveta Mikhailovna. "Nuclear Energy Status in Russia: Historical and Contemporary Perspectives." *Journal of Humanities and Social Science* 21, no. 9 (September 2016): 37–46.

Srinivasan, M.R., R.B. Grover, and S.A. Bhardwaj. "Nuclear Power in India: Winds of Change." *Economic and Political Weekly* 40, no. 49 (2005): 5183–5188. www.jstor.org/stable/4417490.

Stocker, T.F., ed. *Climate Change 2013: The Physical Science Basis. Contribution of Working Group1 to the Fifth Assessment Report of the intergovernmental Panel on Climate Change*. Cambridge, UK: Cambridge University Press, 2013.

Wolfe, Audra J. *Competing with the Soviets: Science, Technology, and the State in Cold War America.* Baltimore, MD: The Johns Hopkins University Press, 2013.

Xu, Y.C. "The Struggle for Safe Nuclear Expansion in China." *Energy Policy* 73 (June 16, 2014): 21–29. doi: 10.1016/j.enpol.2014.05.045.

Yang, Jin, Donghui Zhang, Mi Xu, and Jinying Li. "China's Nuclear Power Goals Surge Ahead." *Science* 340 (April 2013). doi: 142. 0.1126/science.340.6129.142-a.

Part III

The renewable energy regime

7 Hydropower

Four case studies

Introduction

During the twentieth century, governments and private utilities have transformed their relationship with rivers and their tributaries by building as many as 45,000 large hydroelectric dams. Proponents argued that with electricity usage projected to nearly double by 2035 from 5.2 terawatt-hours to 9.3 terawatt-hours, big projects such as large hydroelectric dams offered many benefits.[1] In addition to meeting electrical needs, the benefits, according to proponents, included a reduction in fossil fuel consumption, flood control, irrigation, improved water transportation and employment for the construction industry. After 1945, developing countries, with aid from the World Bank and the International Monetary Fund, dams producing electricity became fixtures on nearly all of the world's largest rivers. On the negative side, as many as 60 percent of the dams and reservoirs financed in this way removed between 30 and 60 million people from their homes.[2]

In recent years, the momentum to build more large dams to produce electricity has slowed in the developed world as evidence accumulates, suggesting that large dams as sources of clean energy do not leave a light footprint on the land. There now exists extensive documentation of the ecological damage done to the land and to aquatic life by large dams. For example, governments and utilities removed 538 dams in the 90 years before 2005. In eight years from 2006 and 2014, they removed 548 dams, many of them considered unsafe or replaceable by more efficient means for producing electricity. In 2015, Washington State in the United States removed two large dams, one 64 meters high (210 ft) and the other at 32 meters (105 ft), releasing 10 million cubic meters of sediment into the rivers.[3]

During the past century, countries with large flowing rivers, Brazil, China, Canada and the United States have become the largest producers of hydropower. This chapter explores the development and power production of large dams in these countries, in Brazil, with rivers like the great Amazon and Parana; in Canada, with the Yukon, Nelson and Columbia Rivers; in China with the Chang Jiang (Yangtze) and the Huang He (Yellow) Rivers; and in the United States, the Missouri and the Mississippi Rivers.

The reasons for exploring some of the world's largest rivers include their discharge and length. They begin in high terrain and flow to the open sea.

Mountains created by tectonic uplift usually provide the momentum needed to create a river's flow. However, the source of the Mississippi River in the United States is Lake Itasca in Minnesota. It has an elevation only 461 meters (1,515 ft) above sea level. For most rivers, however, the flow's kinetic energy sends debris and sediment downstream. A rapid flow cuts into the terrain around it and reshapes the landscape. For example, over millennia, the Colorado River eroded the land around it and created the picturesque Grand Canyon. Sadly, damming the Colorado to provide hydropower for the growing populations of Los Vegas, Los Angeles and many other Arizona and California cities prevented the river from reaching the Gulf of California and depositing its sediment. There, its much-needed sediment created a delta for aquatic life located in a climate where temperatures exceed 100 degrees or more during summer months. Instead of reaching the Gulf, this overused river, dammed at Glen Canyon and the Grand Canyon, among others, dries up in the Mexican desert.

Large dams have also altered the relationship between humans and the world's rivers and their basins. A combination of scientific and engineering expertise, political power and the economics of development have directly displaced between 30 and 60 million people.[4] Newly created reservoirs altered the lives of those living downstream by disrupting their social networks and their livelihoods dependent on fish and other aquatic life. Flooding upstream communities for reservoirs resulted in the forced removal of 500 million people.[5]

Unpredictability described more accurately the impact of large dams on those most affected by their construction. Consumers living at a distance reaped the benefits of new or enhanced electricity:

> [T]he lessons of history are that large dams and river basin planning are complex hybrids of nature, technology, and society. These hybrids behave in often unpredictable ways, despite the best efforts to plan for and take account of the social and biophysical changes wrought by damming a river.[6]

Brazil

Dam building peaked in 1968 throughout Western Europe, the United States and Canada. Not unlike the United States, Brazil's growing population, possibly unsustainable using number metrics, is concentrated unevenly in coastal areas. The availability of its more than 3,000 rivers, representing 17 percent of the world's freshwater, made hydroelectric power development a natural for development. Development of this major energy source ensured the availability of enough hydroelectric power for domestic, industrial and agricultural growth. With more than 60 percent of its energy supply coming from the hydropower generated by more than 600 dams, Brazil is most dependent on this energy source. Making hydropower sustainable and protecting its environment from dam building's overreach and protecting the rights of the country's indigenous populations has become a major societal challenge.

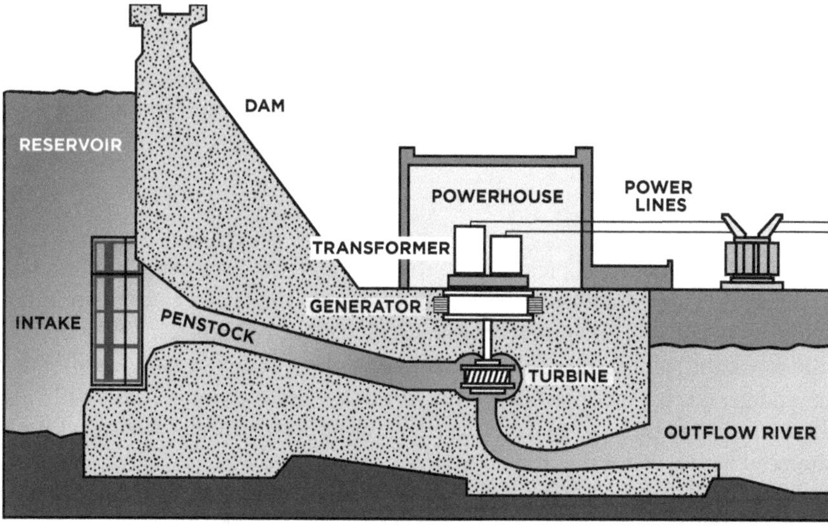

Figure 7.1 Diagram of a hydroelectric power plant.

Since the 1970s, a time when dam building in the developed world ended, the opposite became a reality in South America and especially in Brazil. For the continent, the building of two new large dams every three years became the norm with Brazil leading the way. This new normal is projected to continue during the present decade and beyond. Much data support this forecast. South American countries possess 20 percent of the world's exploitable hydropower, with Brazil dominating the other countries on the continent. To date, large dams capture only 10 percent of its energy potential to produce electricity.[7]

Of the major South American rivers, the Parana River in southeastern Brazil has attracted the most intense development of large dams and will remain so for decades to come. The Parana's headwater tributaries, the Paranaiba, Grande, Tiete, Paranapanma and Iguazu Rivers, will receive the most exploitation for generating hydropower in areas with the greatest population growth. In addition to hydropower, dams store and distribute water for irrigation and flood control. Brazilian dams, like others around the world, alter riverine downstream ecosystems. Dams negatively affect the migration of fish and the water temperatures they need to thrive rather than to merely survive. Damming a large river compromises its channel morphology. It includes the distribution of aquatic plants and changes in the connectedness of a river's alluvial environment, including its floodplains, its wetlands and forests.

Large Brazilian alluvial rivers share common characteristics as they flow downstream. At midstream, they spread out a considerable distance reaching out to several smaller rivers. As a result, they form large internal deltas. The deltas of large Brazilian rivers, especially the Amazon and Parana Rivers, maximize their biological diversity serving as spawning and feeding areas for a variety of fish habitats. In the main channel of these two rivers, freshwater dolphins, turtles and crocodiles thrive by taking refuge during droughts. Few species use the main channel exclusively, and those that do tend to be predators in search of prey who depend on floodplain habitats.[8]

The Amazon biome

As a highly diversified tropical biome, the river and rainforest serve as the home to the world's largest collection of living species. As the largest river basin in the world, its length of 4,195 miles, covering 2,720,000 square miles, includes 15,000 tributaries and sub-tributaries.[9] Along the main channel and its many smaller rivers, indigenous populations numbering in the hundreds of thousands and living there since the last Ice Age have waged centuries-long struggles to protect their ancestral homes and ways of life.

When the Portuguese explorer, Pedro Alvares Cabral arrived in the region in 1500 CE, an estimated 2,000 indigenous tribes inhabited the biome that became Brazil. From that beginning, the biome's abundant resources became a source of conflict. Immigrants came from Portugal seeking a better life. Banks, large corporations and private investors came to exploit the biome's riches in gold, brazilwood and aquatic species. West African men and women captured by Portuguese shippers and sold as slaves to plantation owners brought skills as potters, weavers, boatbuilders and culinary innovators and much more to the Amazon. Along with slavery, Europeans brought nonnative diseases that decimated indigenous populations; tribes in the thousands became tribes in the hundreds. Those tribes that remained faced a steady onslaught from rubber tappers, loggers, plantation owners and peasant farmers.

With about 80 percent of Brazil's electrical energy coming from hydropower, the incentive to build larger and more efficient dams became a national priority. Economic and population growth provides the rationale for more dams. However, dam building on Brazil's major rivers in the last decades has resulted in a concerted effort to delay and ultimately prevent dam construction.

The Xingu controversy[10]

The damming of the Xingu River in Northwest Brazil with the Belo Monte Dam represented just one of the major large dam projects in the first decade of the twenty-first century. Odebrecht, a Brazilian engineering and construction conglomerate, the government of Brazil and the Norte Energy Consortium a private contractor planned Belo Monte as the world's third-largest hydroelectric dam, with the Three Gorges in China holding first place followed by Canada's Syncrude Tailings Dam.

With projected costs of US$19 billion and generating 11,233 megawatts of electricity, Belo Monte was planned to begin operation in 2015. As planned, the complex included two dams, two artificial canals, two reservoirs and a series of intricate dikes. In the process, Belo Monte would flood 500 square kilometers of rainforest, raising havoc on the surrounding natural environment and displacing approximately 20,000 indigenous persons and isolated tribes. One of the proposed construction's most apparent ecological impacts with both human and environmental consequences would be changes in the water chemistry of the Xingu River. The higher water temperatures in reservoirs change to cooler cleaner water temperatures as it passes through the dam's turbines to generate electricity. Alterations in water oxygen adversely affect the river's aquatic life, including all species of fish. They have evolved to thrive in waters with traditional Amazon temperatures rather than those created by artificial means. River dolphins face the same altered environment and suffer accordingly.

These environmental changes, including water chemistry and temperatures, threaten the survival of indigenous populations who depend on the river's aquatic life for much of their food, fish being the primary food source. According to Amazon Watch, a nonprofit organization founded in 1996 to protect the rainforest and the rights of indigenous people, the damming and diversion of the Xingu River will leave many tribes, including the Yudja, Arara, Kayapo and the Jorunas, without water, fish and the means of river transport.

During the 1970s, Brazil's military junta trampled on the rights of tens of thousands indigenous people. By 2007, such action would be in violation of the United Nations Declaration on the Rights of Indigenous Peoples, signed by most member countries, except the United States, Canada, New Zealand and Australia, countries with problematic relationships with its First Nation people. At the end of the 1970s, the overthrow of military rule and its replacement with a democratically elected government created a more transparent society in Brazil. Indigenous groups gained access to the government's dam-building plans to organize and join other groups with similar grievances against Brazil's plans for economic growth based on hydropower. Protests served as a catalyst against dam building in rural Brazil.[11]

The planned Xingu River Basin Hydroelectric Project, a huge six dam project that would flood thousands of square kilometers of tribal land brought together rural Brazilians who suffered greatly by previous hydroelectric projects and received little or no compensation for their loses. As one tribal chief explained it, regarding the first phase from 1974–1984 building of the Tucurui Dam "They said that they would compensate us but Eletronorte blocked our claim in court. You can't trust them. They say they are conducting studies. They told us that. But with each study, they sealed our fate. Little by little, they moved in. Then the dam was built."[12]

A television advertisement produced in 1987 brought the world's attention to the plight of Brazil's rural indigenous populations. In it, a Kayapo tribal woman wielding a menacing machete over the head of one of Eletronorte's chief engineers proclaimed, "We don't need electricity. Electricity won't give us food. We need the rivers to flow freely: our future depends on it. We need our forests to hunt and gather in. We don't want your dam."[13] The image in the advertisement went viral attracting the attention of international environmental organizations. With their help, Brazilian activists opposing the planned Xingu River Basin Project traveled to Washington D.C. to meet with government officials and members of the World Bank. Among its many initiatives in the developing world, the Bank funds hydropower projects. Not pleased, to say the least, with an international development agency and the U.S. government welcoming feedback on the Xingu Basin Project, officials in the Brazilian government lashed out at the Basin Project's opponents as meddling in the country's internal affairs.

The charge of meddling backfired badly for the government and led to a television commentator to ask dismissively, "If they're foreigners, what are we?"[14] In Europe, environmental justice non-governmental organizations (NGOs) along with indigenous rights groups protested the project. They lobbied the World Bank and demonstrated at Brazilian embassies. In response, the World Bank withdrew its proposed loan of one half-billion-dollars to the Brazilian Government in March 1989 to partially fund the Xingu River Basin Hydroelectric Project.

The Xingu River Basin Project represented a culmination of a sustained effort to thwart government energy policy, that is "the rapid expansion of electricity generation and electricity-intensive industries through the building of hydro mega-projects."[15] Opposition to these massive projects did not, however, begin with the Xingu. A decade before beginning in 1977, the utility, Eletrosul announced plans to build 22 dams in southern Brazil on the Uruguay River and its tributaries.

Opposition came swiftly from trade unionists, religious groups, land reformers and small farmers. As opposition solidified, a new NGO was created to speak with one coherent voice for Brazil's rural poor. The Regional Commission of People Affected by Dams forced Eletrosul to bargain with this new entity and establish new guidelines for compensating the displaced Brazilians. No longer would the utility be allowed to offer cash to only those who held the formal title to the land. In addition, evicted farmers would no longer accept colonization schemes in the Amazon rainforest that willfully placed them on unproductive land.

With the Regional Commission representing the rural poor, Eletrosul agreed to buy land from large ranchers, pay for the needed infrastructure, including roads, bridges and resettlement schemes. Once completed, the new settlements would grant land to the rural landless poor and to displaced farmers. The 1,620-megawatt-hour Ita Dam was scheduled for 1992. By 1995, preliminary construction only had begun with only a few hundred, rather than the estimated 4,000, families resettled. Committees established by the Regional Commission rather than individuals served as negotiators for the displaced. Power in numbers at the bargaining table proved to be a successful strategy. By making agreements for fair compensation, Eletrosul discovered that its costs exceeded the budget for building the 1,200-megawatt-hour Machadinho Dam. Because of continued protests, the construction site was moved upriver on the Pelotas in 1988. On February 16, 2002, the first generator owned and operated by Machadinho Energetica began to operate and, by August 31, 2002, generated 1,450 megawatt-hours of electric energy. Many dam projects that followed government and private power

Figure 7.2 Demonstration in Brazil against the building Pare Belo Monte power plant.

producers in the name of development, progress, and modernization trampled on the rights of indigenous rural Brazilians.

Does Brazil need large dams?

The obvious answer is, it depends. The government and private utilities have developed the capacity by generating about 77 percent of the country's hydroelectricity. As many as 157 hydroelectric dams with power higher than 30,000 kilowatt-hours, deliver a steady flow of renewable energy. All toll, current electric capacity or about 74,000 megawatt-hours represents only about 28.4 percent of the total potential of 2.601 thousand megawatt-hours. Much of the remaining potential of 50 percent is located in the Amazon Basin on the Tocantins, Araguaia, Xingu and Tapajos Rivers. However, priorities other than hydropower exist in the basin, and they include protecting the rights of indigenous people, fishing rights, irrigation, tourism, cultural diversity and biogeography.

Many physical–chemical–biological problems have led expert observers to conclude that hydropower dam projects become unsustainable over time. Although proponents of hydropower view it as a form of clean energy production when compared to burning fossil fuels, it possesses characteristics that forecast its relatively short life. Arbitrarily, dams inhibit the kinetic energy of rushing water and create reservoirs holding back millions of cubic feet of water. In the long-run dams, however, high and deep succumb to the pressure of holding water and millions of tons of sediment. Hydrologists place the short life of most hydroelectric enterprises at fewer than 100 years. In addition, their accumulated weight on the soil and subsoil creates the conditions for earthquakes.

Dams alter the hydrological cycle of naturally flowing rivers disrupting downstream activities. Still water behind the dams reduces water quality, raises its temperature and delays the natural decay of wastes and effluent. Reservoirs flood the land and accelerate the process of deforestation by destroying the world's photosynthetic capacity to turn carbon dioxide into oxygen. Greenhouse gas emissions, specifically methane, from decaying vegetation found in reservoirs contribute to climate change. Finally, still waters promote the spread of disease vectors that compromise public health. With hydropower being sold to the public as a form of renewable clean energy to promote modernization and progress, hydropower installations need to be weighed against its negative outcomes. In Brazil, the forced displacement of its indigenous population adds another human cost to the claim that such projects benefit the public interest.[16]

The People's Republic of China

Although burning coal dominates China's energy structure presently, the country's undeveloped hydro resources may contribute greatly to China's monumental need for more electricity. The country possesses thousands of rivers with more than 50,000 covering a basin area of more than 100 square kilometers, or 38.6 square miles. Once dammed, 3,886 rivers have a hydropower potential of 10 megawatt-hours of electricity.[17] Currently, China's coal reserves rank first in the world and generate about 80 percent of its electrical capacity while hydropower occupies 15 percent. Fully developed, hydroelectricity will generate about 36.5 percent of China's needs. Petroleum and natural gas contribute 3.4 percent and 1.3 percent, respectively.[18] The worldwide exploration of natural gas in shale deposits may change this distribution. However, as noted in a previous chapter, known natural gas deposits in China may be as plentiful as those in other parts of the world but not as easily accessible. Despite the low percentage for hydropower, China's population and electrical needs rank it first in the world for hydro's installed capacity to generate electricity. However, it lagged significantly behind Brazil and Canada, whose installed capacity produced 83.7 percent and 57.9 percent of their electricity, respectively.[19]

The early history of Chinese hydropower

China's dam building extends over a long historical period beginning about 2,600 years ago. Most were small earthen dams, nothing like the modest 15-meter (42-ft)-high dams of the modern era. Many of these early dams provided the irrigation needed to grow crops and control floods to prevent precious agricultural land from being washed away. Some continue to provide these functions today.

The history of dam building by the Chinese government to produce 500 kilowatt-hours of energy began in 1919. By then, the country's unifying

Nationalist leader, Sun Yat-Sen identified the Yangtze Gorge, a location that in the last decades of the twentieth century would become synonymous with hydropower and flood control in China. Situated near the city of Ichang in the province of Hubei, this wild river, sometimes described as a place of "spectacular and sometimes violent beauty,"[20] descends from mountainous headwaters cutting deep canyons or gorges in the terrain. The Chang Jiang (Yangtze) also replenished its floodplains over the centuries with sediment. They became fields of cultivated wheat and other food staples for an impoverished population lacking social services and adequate transportation.

Catastrophic events delayed China's efforts at modernization. Japan's aggressive expansionist policies led to its military invasion of China's Manchurian Province in the 1931–1932. Japanese expansion into mainland China, Southeast Asia and the Pacific Islands led to its attack on the U.S. naval base at Pearl Harbor in the Hawaiian Islands on December 7, 1941. It brought the United States into a conflict that would become the Pacific theater of World War II (1941–1945). The war temporarily halted a civil war between the Chinese Nationalist government led by its United States ally, Chiang Kai-shek and the Communist insurgents led by Mao Zedong. The hiatus during the war years allowed the Nationalists and the Communists to focus their efforts on defeating Japan.

Recognizing the geopolitical ramifications of a Japanese victory, the United States provided logistical military and economic assistance to Chiang's Nationalist government. It believed that "the peace of the world [was] tied up with China's ability to win or prolong its resistance to Japanese aggression."[21] Infrastructure development became one way to shore up and stabilize a government trying to defeat an external aggressor and win a civil war. The Yangtze-Gorge Project became the centerpiece for China's modernization.

With support from U.S. officials across the spectrum of federal agencies, including the Bureau of Reclamation and its embassy, the Nationalist government weakened by Japan's conquest of major sections of the country forged ahead with plans for a major infrastructure project. It would be "a massive straight concrete gravity dam roughly 225 meters in height." Twenty power and diversion tunnels would be built on either side of the river. Its power plants would produce 10,560 megawatt-hours of energy, a feat unknown in the world during the war years.[22] The aftermath of the war and a communist victory in 1949 scuttled the project. "Had the project gone ahead, it would have easily set records as the most expensive and massive engineering feat on the plant."[23]

The civil war had begun in 1931, delayed through the war years (1941–45) and reignited soon after the war's end with Japan's unconditional surrender. Despite the hiatus, the civil war lasted from 1931 to 1949 and ended in victory for the communists and the retreat of the Nationalist forces to the island of Formosa (Taiwan). The communist victory in the long and protracted civil war ended U.S. plans for the modernization of China. The proposal that Sun Yat-Sen made in 1919 captured Mao Zedong's attention, however. He swam with comrades in the river, wrote a poem celebrating the Chang Jiang's (Yangtze's) potential to capture the river's energy. However, his highly destructive policies, including the

Great Leap Forward and the Cultural Revolution, that followed drove the country into an economic and psychological crisis. Not until 1994, many years after Mao's death, did construction commence.

Since the establishment of the People's Republic of China in 1949, four surveys of the country's hydropower needs have resulted in the acceleration of dam construction. Rapid industrial development, the modernization of older cities and the building of new ones became government priorities, all requiring a significant electrical capacity. Today, large dams provide 145.2 gigawatts and 486.7 terawatt-hours (10^{12} watt-hours). The country's economic plan to modernize with coal and hydropower is intended to provide electricity to the increasing urban population. Hydropower's contribution grows with the construction of 22,000 dams since 1950. Currently, the plan includes the building of dams "in great staircases, with reservoir upon reservoir-some 130 in China's Southwest. By 2020, China aims to generate 120,000 megawatt-hours of renewable energy, most of it from hydroelectric power."[24]

Three Gorge Dam

The Three Gorges Dam (TGD) is the largest such power plant in the world. It is composed of three integrated construction projects. The key includes the dam, its powerhouses composed of 26 hydro-turbines, generating 700 megawatt-hours of electricity each, the ship lock and ship lift. The reservoir, with a normal level of 175 meters (575 ft), with a total capacity of 39.3 billion cubic meters, and the transmission projects make the dam functional. Unlike most of the world's dams that provide irrigation for the surrounding land, this dam was designed with many objectives in mind. Fewer than 25 percent of the world's dams produce hydropower. Yet, the TGD's primary objective was to produce electricity, upward of 18.2 gigawatt-hours (billion watts) annually. By storing water in its massive reservoir, it was intended to control the flow and prevent downstream flooding.

Partial completion in June 2003 of what became known as the Three Gorges Dam began its first stage by filling the reservoir with 445 feet (136 m) of water. Stages two and three brought the level to 575 feet (175 m). To place the impact of this reservoir in context, at 575 feet high, it created a lake 410 miles (660 km) long. It exceeded the length of Lake Superior, the largest of the Great Lakes in the United States by 60 miles (97 km). Since TGD mirrored the narrowing of the Chang Jiang, the lake/reservoir inundated the land previously inhabited by 1.3 million Chinese peasants who lived in 13 cities, 140 towns and 1,600 villages.[25] Controlling water flow, its proponents argued would make the river more navigable for ships to travel through the Gorges to cities upstream. A series of locks designed to lift or lower ships on their upward journey would provide a boon to economic development in the region. Renewable energy would replace the airborne pollution caused by China's massive coal reserves. At what costs would these goals become achievable?

Figure 7.3 Three Gorges Dam across the Chang Jiang (Yangtze) River.

Geological and ecological effects of the three rivers gorges dam

According to much of the scientific research regarding the planetary impact of massive diversion dam projects around the world, shifting and slowing the flow of the world's largest rivers, slowed Earth's rotation. TGD on the Chang Jiang River, one of the most debated hydraulic projects in the world, may accelerate this process with unknown planetary effects. The government claims that its dam-building initiative is safe, will avoid pollution, will respond to the global climate change plans to reduce carbon emissions, will control floods and droughts and will improve the quality of life for its citizens. Many scientists in China regarded the TGD project among the most contentious undertakings by the government's hydraulic development plans.

The effect of collecting billions of tons of formerly flowing water in reservoirs behind TGD has the potential of fracturing Earth's fragile mantle. Since the weight of the reservoir sits atop two tectonic faults, the Jiuwanxi and the Zigui-Badong, its water level rises and falls with floods and drought. Such variability strains the mechanics of these faults and induces earthquakes. Called reservoir-induced seismicity, geologists believe that the tremors caused by earthquakes have their origin in dam building. Chinese engineers believe that at least 19 earthquakes in the last 50 years originate in seismically active zones. Some caused minor tremors while at least one near the province of Guangdong's Xinfengjiang Dam in 1962 measured 6.1 on the Richter scale and caused major residential destruction in a heavily populated area.[26] Scientists believe that similar seismic events followed the completion of TD with 822 tremors registered after the dam's reservoir reached its capacity of 575 feet in 2009.

Landslide activity increased along the river's path, its reservoirs and the dam caused by the increasing pressure placed on the surrounding terrain. The loss of the river's aquatic diversity and the destruction of its flora and fauna along now-flooded land coupled with the resettlement of millions of people became victims of China's modernization. Biodiversity in all of its life forms, flora, fauna, and humanity became victims of the compulsion to modernize with electricity and to see the world not as stasis but as movement and as progress for achieving a goal or a number of goals. As noted in the case of the Amazon Basin, one of the world's most diverse biomes, hydropower brought electricity to many but at a high ecological price. It displaced and required the resettlement of millions of indigenous people. It transformed the land and the fauna that thrive in its diverse habitat. It caused an equally disruptive impact on the aquatic life in the upstream rivers and the basins that empty into the sea.

Deforestation became widespread throughout China's central provinces once flooding of the Three Gorges reservoir became a reality. Along with the flooding came waterborne pollution as runoff from the surrounding terrain flowed into the reservoir. It endangered plant species, including China's dove tree and its dawn redwood. According to one ecologist Jianguo Liu, as many as 400 plant species were threatened with extinction. Inundated vegetation, including trees and plants rotting at the reservoirs' bottom emit carbon dioxide and methane once they reach the surface. When water levels drop, these greenhouse gasses reach the surface more quickly.

The construction phase also produced greenhouse gasses with the transporting of materials and operating heavy equipment used in construction. The production of cement and the manufacturing of steel led the way in carbon dioxide emissions. Reductions in carbon dioxide aside, the efficacy of building large dams remains debatable. Their construction costs remain prohibitively high given reports that they operate at 80 percent of their targeted goal in the first years of operation and too high to yield a positive return. The construction of large dams takes years to complete, so they remain ineffective in meeting immediate energy needs. Mitigation costs for environmental damage reach tens of millions of dollars. Those for TGD in a 10-year period are projected to cost US$26.45 billion.[27] Once completed, the dam's reservoirs, and this is not restricted to the Gorges, become receptacles for the runoff of chemicals and fertilizers. Human and animal waste, as well as all kinds of trash, seep into reservoirs. "During the 2010 flood, floating refuse backed up behind the Three Rivers Gorge over an area of more than 50,000 m^2, so thick that people can literally walk on the water's surface."[28]

As the decreased flow of the Chang Jiang empties into the South China Sea, it disrupted the habitat of the river's 177 fish species, already compromised by decades of overfishing. With decreased flooding downstream, the network of depleted lakes where most of the fish evolved and live are placed on the path to extinction. Unable to migrate and breed, river fish stocks collapse. Ecologists considered the lowered water levels as a primary cause of depleted fish stocks.[29]

By reversing the natural flow of rivers, by storing water in wet seasons and releasing it during droughts, marshes and wetlands dry out, making it impossible for the land to absorb the high volumes of water. With a warming climate, some regions of the world will experience more intense rain while others will trend in a drier direction. Regions that depend on glacial melt will witness a diminished flow. Reservoirs will not fill to capacity. Dams that deliver the reduced flow to their many turbines to generate electricity will fail to meet their production goals.

Canada

Canada is a country of wild rivers, with the Mackenzie, the Fraser and the Churchill among its wildest and longest. It is also a country determined to tame its rivers for hydropower. Almost 60 percent of its electricity comes from dams. By comparison, dams produce only about 16 percent of the world's electricity. This case study focuses on the dominance of Hydro-Quebec in the hydropower industry. As the largest electric utility in Canada and one of the largest in North America, it owns 49 hydroelectric dams, with 93 percent of its hydropower coming from them. British Columbia Hydro and Manitoba Hydro, not subjects of this case study, follow as Canada's largest hydropower providers.[30]

In 1885, Canadian engineers built its first Hydro-Québec power station. They dammed the Montmorency River at its impressive waterfalls, almost 100 feet higher than Niagara Falls. At the time, an intense competition developed among traditional gas and the new electric light companies. To the amassment of an assembled crowd, the Quebec Levis Electric Light Company's power station at Montmorency Falls transmitted electric power to 34 electric arc lamps to Quebec City's Dufferin Terrace. This event marked the beginning of electric power transmission in North America.[31] Population growth, industrial development, irrigation and flood control over the next century led to the construction of 48 dams on the river and its tributaries. With technological improvements, authorities decommissioned the original river dam and replaced it not far upstream by the Marches-Naturelles Hydro-electric Power Station with an installed capacity of 4.16 megawatt-hours.

By 1889, the Royal Electric Company in Montreal eclipsed the city's gas companies by winning the public lighting contract for the city. In 1901, it merged with the Montreal Gas Company to create Montreal Light, Heat and Power Company (MLH&P). The replacement of horse-drawn streetcars with those powered by electricity followed the illumination of its streets with hydropower. In 1897, a private utility built a generating station in Sherbrooke, Quebec for both electric lighting and transit. Other cities in the province followed this innovation quickly.

During the 1920s, capturing the power of Quebec's many rivers accelerated with the construction of 80 generating stations providing electricity to cities, industries and agriculture. In the beginning of what became a boomtown mentality soon led to the elimination of competitors and the consolidation of the many utilities into powerful regional monopolies. Cost-cutting MLH&P eliminated the city's many private utilities in a war over electricity rates. Its preeminent position in the energy market gave it the power to refuse cooperation with provincial commissions and agencies established to regulate costs.

In Quebec's Maurice region, a similar pattern unfolded. There, the Shawinigan Water and Power Company eliminated competitors and built an impressive industrial complex, despite the absence of a large population to buy the hydroelectric power generated along the Riviere Saint-Maurice. It engaged in a strategic plan to sell its electricity. It targeted the paper, pulp, aluminum and chemical companies to buy its electricity by using a pricing system that reduced costs with high volume purchases.

For households, a traveling road show extolled the virtues of the new kitchen, with its electric appliances – stoves, refrigerators, coffee makers, toasters and the like. It encouraged industrial development along its waterways by investing in companies. It began the practice, later used by other utilities, to expand its market by exporting some of its power to distant markets. In 1903, it built a 50-kilowatt-hour voltage, 75-mile-long transmission line on wooden poles beyond Quebec Province. Over a 50-year period, eight power stations harnessed this mighty river's kinetic power.

Hydro-Quebec

The global Great Depression (1929–1938) brought many private utilities to the brink of bankruptcy while a number of companies whose stock was traded on the exchanges of countries around the world experienced such a drop in value that they either faced bankruptcy or a severe decline in productivity. One of the beneficiaries of this global event was the mighty MLH&P that purchased failing utilities. When the government of Ontario Province canceled a contract to purchase power from one of its largest providers, Montreal Light bought Beauharnois Light, Heat and Power, a failing utility. It had begun construction of a massive hydropower project on the St. Lawrence River.

Dredging the Beauharnois headrace canal, 3 miles wide, 70 miles long and 30 feet deep was like dredging of the Panama Canal. Completion took place in stages with the last power generating station, number 36 becoming operational in 1961. It ended more than 30 years of construction. The history before nationalization was characterized by antagonistic confrontations between electricity companies and the Canadian government. The government responded to complaints from the public that companies charged exorbitant electricity rates made excessive profits, provided poor service and remained belligerent as government agencies attempted to regulate rates and pricing of the utility companies. Reports commissioned by the Canadian government attempted to impose regulations without success. So, in 1944, during the height of World War II, the government expropriated the financial assets of the privately held electricity and gas companies that had monopolized the industry and refused to yield to government regulation. A provincial corporation, the Quebec Hydro-Electric Commission took control of these assets and on April 14, 1944, creating Hydro-Quebec. At the time, this new public corporation inherited four generating stations and a transmission network, a mere infant enterprise when compared to its growth in the decades that followed. To compensate the shareholders of the private utilities, the commission initially took out a bank loan that was eventually replaced by a 12-million-dollar bond issue that paid a quarterly dividend to buyers.

In the 1950s, this new behemoth extended its reach throughout the provinces of Quebec and Ontario. It constructed more power to the Quebec Cote-Nord region of the province by laying an underwater cable to the Gaspe Peninsula and on the lower portions of the Ottawa River in Ontario. Initially, the central government constrained Hydro-Quebec's growth by leaving local rural communities with the responsibility of providing their own electricity. The passage of the Rural Electrification Act in 1945 affirmed this initiative.

A total of 46 electricity cooperatives sprung up in small towns in response to the law. However, despite government subsidies, construction and maintenance costs became burdensome over time. Many of these rural areas contained riches in the form of timber and mineral resources, all of which required electricity. As a result, by 1963, all but one cooperative accepted Hydro-Quebec's offer to become part of its growing electrical grid. Only Saint-Jean-Baptiste-de-Rouville remained independent. Hydro-Quebec's reach extended more than 1,800 miles and became the first utility in the world to use 315- kilowatt-hour transmission lines. Technological breakthroughs accelerated as this massive public utility reached further afield, providing hydropower to more communities. It made a significant breakthrough in long-distance transmission by raising the voltage to a record level of 735 kilowatts. Despite Hydro-Quebec's success, the electrical industry at large continued to be plagued by inefficiencies in service, disparities in electrical rates and overlapping responsibilities. Private distributors, electric cooperatives and cities with their own power stations continued generating electrical power for their own uses. Companies, especially Alcan, sited its aluminum plants along rivers to gain access to the water's kinetic energy.

The provincial government and voters approved a second nationalization of power generation for Quebec. Over a three-year period, Hydro-Quebec acquired 80 companies, private distributors, cooperatives and municipal holdings. Uniformity became the standard for all electrical rates and for transmission networks. To reimburse shareholders for this second nationalization, Hydro-Quebec sold US$300 million in bonds to investors in the United States. By doing so, it made this complex and unprecedented merger into a hydropower behemoth. Today, Hydro-Quebec, a nationalized public utility, operates the dams on the Saint Lawrence River. Its massive market for hydropower extends beyond the province to Ontario and New England and remains Canada's largest power producer and one of the world's largest and most powerful generating stations.

The James Bay Project

With James Bay located just south of Hudson's Bay, it became one of Hydro-Quebec's largest and most controversial hydropower projects. Indigenous

Canadians rejected its destructive impact on their communities that depended almost exclusively for sustenance by hunting for moose and beaver and for fishing in its waterways. They comprised three distinct groups: Inuit, who were granted control of Northwest Territory in April 1999 and renamed it Nunavut; Metis; and First Nations. This later group comprises as many as 50 language groups with different histories, languages and spiritual beliefs. At present, aboriginal people represent 5 percent of the population and control 20 percent of the land. With its population growing faster than Canada's European and Asian descendants, clashes over dam building became predictable. Hydro-Quebec and the Cree, Inuit and Naskapis indigenous people settled the first stage of the James Bay Project amicably.

They received exclusive rights to a portion of the unused land and financial compensation of CN$225 million. Hydro-Quebec's more expansive plans became a source of bitter conflict. Its location on the Great Whale River east of Hudson's Bay would disrupt a vast, fragile ecosystem of taiga composed of black spruce and great conifer forests. This vast expanse dotted by lakes, bogs and ponds came alive during the autumn and spring as giant herds of caribou raced across the taiga to fed, mate and prepare themselves for winter. At the same time, millions of migratory birds on their way to the farthest reaches of South America stopped to feast during a "rite of gluttony" on eelgrass and coastal shrimp, doubling their weight for the long trip across the Americas. A National Audubon Society report noted that migratory birds "may have no substitute habitat if key sites in James Bay are destroyed."[32] Furthermore, the study concluded, "many species of migratory shore birds would be severely threatened, possibly even to extinction, if crucial stopover areas were damaged."[33]

Submerging the Great Whale River's basin, east of James Bay and southeast of Hudson's Bay, with a vast dam and reservoir system, would forever alter this landscape and permanently disrupt the migratory patterns of caribou and birds. As planned, the project would become one of the world's largest hydroelectric complexes with water from its dams churning turbines many times the force of those at Niagara Falls. The initial stage, south of the river, used enough concrete to build 80 Great Pyramids of Cheops and changed the flow of the La Grange River complex. This part of the James Bay plan represented one of the world's largest construction projects. This plan diverted the river complex and caused it to flow west into James Bay rather than northeast into Ungava Bay. As engineering projects go, it seemed uneventful until 1984 when 10,000 migratory caribou died in the raging waters caused by the controlled releases from reservoirs built to divert water from this river complex.[34]

Further planning was larger in size and ecological impact. With the memory of 10,000 dead caribou, opposition from the 12,000 Cree and the 6,000 Inuit Eskimos became vigorous. This planning phase would create seven new reservoirs. By diverting many of the affected rivers to create the reservoirs, the lower stretches of the rivers would become dry bedrock. The project's impact on the fish cultures of the affected aboriginal communities would be devastating. Political priorities influenced the decision making of Francophone Quebec. It believed that the production of an additional 226,400 megawatt-hours of electricity, 15 percent of which would be sold to New York State, would provide it with the leverage needed to gain independence from English-speaking Canada.

A US$19.5-billion contract from New York would go far to meeting James Bay's financial obligations, measuring some US$50 to US$60 billion. Without such a contract, surplus electricity could not be absorbed by Canada. Despite efforts by Hydro-Quebec to deflect the projected ecological impacts, environmentalists supporting the Cree and Inuit populations concluded, "in terms of wildlife and habitat, the devastation of James Bay is the northern equivalent of the destruction of the tropical rain forest."[35]

Twice, New York State agreed and then chose to reject plans to increase its use of Hydro-Quebec's James Bay power. Protecting the environment and recognizing the rights of the Cree and the state's efforts at energy conservation guided its decision. As protests continued, some of which bordered on violence, the Canadian court system accepted aboriginal claims to the land and rivers. In 1994, Quebec's premier, Jacques Parizeau, indefinitely postponed the Great Whale River project. This edict led Hydro-Quebec to sign an agreement with its adversaries in June 21, 1999. The agreement accepted an environmental review of economically feasible projects. In addition, agreements required the acceptance of local indigenous people.[36]

The saga of the James Bay project and its unwritten final chapters got caught between the promoters of hydropower and recognition of its negative effects on the environment and the social, economic and cultural life of the affected communities. Promoted legitimately as an alternative to burning fossil fuels, renewable energy in the form of an unlimited supply of flowing water became intoxicating. Replace energy produced by hydropower with coal consumption and you require about 200 million tons of coal to produce the same amount of electricity powered by water. "This is equal to a coal unit train of 100 cars every 3 min., for 24h/day and 365 days/yr. moving at a speed of 30 km/h, there would be no gaps between them, and they would become one infinite unit train."[37]

Despite these advantages, the enthusiasm for hydropower as renewable energy faced opposition from some constituencies. First, dam building altered and destroyed the traditional life of hunters and gatherers. Fish, a food staple suffered; mercury released by submerged plant matter contaminated traditional sources of drinking water. By curtailing hunting for caribou, moose and other animals, indigenous communities became dependent on a cash economy. Protest movements with substantial support from others led to court cases, protests at construction sites and some violence.

The concept of growth through ever-increasing developments in energy use conflicted with newer conceptions. Preservation, environmental protection of land and water and the emergence of sustainability to promote energy efficiency and conservation challenged the concept of unlimited and unsustainable growth. It may not be feasible to demand that energy can and must be reduced substantially. However, Canadians and others going forward must understand that hydropower is neither green nor clean. Pouring hundreds of thousands of tons of cement into a free-flowing river also has costs that compare unfavorably with other renewable forms of energy. A substantial case against the continued use of fossil fuels does not inhibit a discussion of hydropower's future as part of a balanced future energy strategy. In the meantime, Canadian energy policy includes the investment of CN$55 to CN$70 billion in new hydroelectric projects, adding 14,500 megawatt-hours to the country's existing 71,000 megawatt-hours in the next 15 years.[38]

The United States

This fourth case study begins with the Appleton Edison Light Company installing the first hydropower generating station on the Fox River in Appleton Wisconsin. H.J. Rogers, a paper manufacturer, impressed by the work of Thomas Edison at his Pearl Street, Manhattan, Illuminating Company financed the project. There, an electrical generating plant provided service within a square mile to the homes and businesses of its 500 customers. Horse-drawn wagons hauled coal to Edison's company. There, large steam engines powered by coal turned the generators to create electricity for its customers. By contrast, the Rogers plant generated hydropower by damming the Fox River and using its turbines to generate electricity, enough to power Rogers's mill, his home and one additional building. It began operation on September 30, 1882, the first of its kind in the world.

Unlike the previous case studies, the greatest construction of hydroelectric dams took place between the 1930s and the 1960s. Then, about one-half of the country's 340 active dams began producing hydroelectric power.[39] Prior to this expansion, however, Nikola Tesla and George Westinghouse built the first hydroelectric plant in 1895 on the U.S. side of Niagara Falls to deliver electricity far beyond the local community. Tesla's polyphase apparatus eliminated "the various kinds of current required by different kinds of lamps and motors. The Niagara-Tesla system generated only one kind of current to be transmitted to places of use and changed to the desired form."[40] By using Tesla's method of alternating current rather Edison's direct current, power was delivered to Pittsburgh Reduction Company (later named Aluminum Company of America) and to Buffalo in 1895. To celebrate Tesla's achievements in Niagara Falls, statues show him standing on top of an alternating-current motor in Canada's Queen Victoria Park and another on the U.S. side of the falls at Goat Island, where he is seated and reading a diagram.

These early plants in the late nineteenth century generated power for lighting homes and factories and replacing gaslights on many of the country's city streets. The invention of the small electric motor for household appliances and manufacturing workshops increased the demand for electrical generation. The end of World War 1 witnessed the standardization of most electric plant designs, with transmission and distribution networks becoming priorities.

Large-dam construction

In 1901, Congress enacted its first Water Power Act to eliminate congressional approval for each power plant, thereby streamlining the approval process. In

1907, the Bureau of Reclamation was separated from the United States Geological Service becoming an agency in the Department of the Interior. It provided water resource management to 17 of the states in the nation's desert West. Its network of hydroelectric dams produced enough power to make the desert habitable. Before reaching that goal, however, power plants at dam sites allowed contractors to use giant shovels to dig diversion canals and power sawmills to cut timbers for cable lines and to lift materials for building the dam site. Electricity on the site allowed work to continue after dark.

Once complete, hydropower driven turbines provided electricity at the dam site and sold its surplus power to local distribution systems that served farms, industries and towns. Cheaply priced electricity helped to offset the cost of the desert West's highest priority, the cost of water for irrigation and domestic use. Phoenix, Arizona, benefited from such cost savings during the construction of Theodore Roosevelt Dam on the Salt River, northeast of Phoenix, in 1909. A 4,500-kilowatt power plant with five generators pumped groundwater to irrigate farms and provide cost-effective electricity to the city.[41]

The Roosevelt Dam became the model for the development of large farms to feed the nation's growing population. Western cities, among the nation's largest, purchased electricity at a fraction of the cost when compared to midwestern and eastern cities. The extraction industry blossomed as copper smelting and surface coal mining became instrumental for the country's industrial development. During World War 1 (1914–1918), reclamation projects added in the war effort. After the United States entered the war in 1917, President Woodrow Wilson (1913–1921) ordered the construction of a munitions plant at Muscle Shoals, Alabama, powered by hydroelectricity from the Wilson Dam. In 1920, the Federal Power Commission began granting licenses to private developers to build dams on the country's major rivers.

During the Great Depression (1929–1938) floods and droughts plagued the Western states. To alleviate the volatility of these extremes, President Herbert Hoover's (1929–1933) decided to engage the federal government in dam construction. He authorized the Boulder Canyon Project on the lower Colorado River in 1931 that ushered in the era of large dam construction. The election of Franklin Roosevelt (1933–1945), during the worst years of the Great Depression, extended federal work projects to relieve massive unemployment. Reclamation, the most visible of these relief projects in the West, resulted in the building of additional large dams.

In his first year as president in 1933, Franklin Roosevelt authorized the building of the Grand Coulee Dam on the Columbia River in Washington State, the Central Valley Project in California, the Tennessee Valley Authority (TVA) and the Bonneville Dam on the Columbia River on the Washington/Oregon border. Government reports, including its "Columbia River and Minor Tributaries," singled out the Columbia River as having the potential for becoming the greatest system for waterpower in the United States. The report recommended the construction of 10 dams on the river, including Grand Coulee and Bonneville, as well as storage dams upstream that in the future would become hydropowered sites including the Hungry Horse Dam. Electrical power received the highest priority, but navigation, flood control and irrigation protected the region's flourishing agricultural development. Over its lifetime, Grand Coulee has produced 1.4 trillion kilowatt-hours of carbon-free electricity.[42]

With the passage of the Public Utility Act in 1935, the Federal Power Commission set wholesale electricity rates while the states regulated consumer rates. Wholesale rates powered industrial growth and prepared the country for the soaring power needs during World War II. The Hoover Dam (Boulder Canyon Project) on the Colorado River became at the time the world's largest hydroelectric plant in 1936 during the height of the country's Great Depression. Dams and those that followed in the American West promoted a surge in population growth, especially in the desert southwest.

As noted earlier, Woodrow Wilson's planned the building of Wilson Dam at Muscle Shoals, Alabama during World War 1. The plan floundered even though Henry Ford offered to purchase the site in 1921 for the purpose of making it a "new Eden of our Mississippi Valley."[43] Opposition from private utilities did not deter Progressives from pushing for an expansive hydroelectric development for the entire Tennessee Valley. Inspired by ideas for regional development proposed by Senator George Norris, (R-NE), Franklin Roosevelt visited Muscle Shoals in 1933. Ideas and the site visit convinced the president to propose a regional program of 30 dams to generate hydropower and control floods. Making fertilizer, combating soil erosion and curtailing deforestation became regional priorities. Turning the tempestuous Tennessee River into an inland waterway for ships would foster regional economic development. To ensure its success, Congress created the TVA, a government-sponsored public corporation, to oversee current and future developments. The TVA became a multipurpose regional program with an educational rationale "geared to changing perceptions of people said to be a hundred years behind the rest of the country."[44]

In accordance with the Colorado River Compact (1922) signed by the seven states in the river's basin (Colorado, New Mexico, Utah, Wyoming, Nevada, Arizona and California), all agreed to a set of water management standards based on the River's flow in 1922. With a changing climate, however, the Colorado's flow, set originally 20.2 cubic kilometers per year in 1922, has a current flow of 16.5 cubic kilometers and declining. As the region's only water source that irrigates and provides hydroelectricity in the "Great American Desert," the Colorado River runs dry before reaching the Pacific Ocean. Increasing periods of drought during the early decades of the twenty-first century may exacerbate current water shortages.

The war years (1941–1945)

Despite the damming of the Columbia, Tennessee and Colorado Rivers in the 1930s, Axis Powers (Germany, Italy and Japan) enemies of the Allied Powers possessed more than three times the electrical power available to the United States upon its entry into World War II. To compensate and catch up, Congress authorized a war budget of US$56 billion to develop a power capacity of 154 billion kilowatt-hours annually to manufacture war material. To meet the President's goal of 60,000 new planes in 1942 required 8.5 billion kilowatt-hours of electric power. Of this total, reclamation projects produced more than 5 billion kilowatt-hours that resulted in a 25 percent increase in the aluminum need to build these planes.[45]

During the war years, these Reclamation projects in the West produced 47 billion kilowatt-hours of electricity. What did this production contribute to the war effort? It was enough to contribute toward the production of 69,000 airplanes, exceeding the president's goal, and 5,000 ships and tanks each, as well the material needed to make these war machines viable. In the postwar period, the aircraft factories, oil refineries, shipyards and chemical companies became lynchpins in a vibrant Western economy.

In addition, most large dams in the United States were built with general revenues from taxes. During the twentieth-century building spree, Congress authorized the Bureau of Reclamation and the Army Corps of Engineers to spend hundreds of billions of dollars from the U.S. Treasury. They built the Bonneville Dam, the TVA and others on the Columbia and Colorado Rivers and many more on the country's navigable rivers. They provided the added power needed during World War II and the economic expansion that followed during the 1950s and 1960s.

The era of building large dams in the United States may be over with the decommissioning of more than 300 large dams during the later decades of the last century. Developing countries and those recently arrived, such as China and Brazil, continue building large dams to modernize their economies and promote upward mobility for their population. In the United States, however, the relicensing of dams requires authorization from the Federal Energy Regulatory Agency. In many cases, the cost of repairs and future maintenance prohibited continuing use. The cost of sediment management in reservoirs accounts for 48 percent of the total decommissioning costs. With most dams, its reservoir storage capacity reaches a limit between 50 and 100 years.[46]

During the recession of the 1970s, federal policy changed regarding dam construction. The government curtailed many subsidies during the presidential administrations of Jimmy Carter (1977–1981) and Ronald Reagan (1981–1989). The Water Resources Development Act of 1986 ended most federal subsidies. With rising costs, state and local governments, aided by private utilities, bore the burden of paying for new projects.[47] Nothing compared to the behemoths produced during the Great Depression and after.

Conclusion

As detailed in the previous case studies, the passion waned for building large dams with government subsidies, except in developing economies where government subsidies and loan guarantees streamlined dam-building projects. In some instances, opposition from displaced indigenous people became highly publicized. Earth Day in the United States on April 22, 1970, revolutionized our thinking about human activities that damaged the environment. The negative effects of building large dams on fast-flowing rivers altered their ecology. Reservoirs inhibited a river's natural flow blocking fish migrations. Since many fish evolved adapting to the river's water temperature, reservoirs raised water temperatures making fish socks vulnerable to disease and death. To prevent sediment from clogging both the turbines and the generators, water emerges from the dam cleaner and colder, damaging aquatic and riparian habitats. Once knowledge about a dam's ecological damage to fisheries, wetlands, water quality and river health became known, enthusiasm changed to caution.

With many small, medium-sized and large dams reaching their maturity of 50 to 100 years, removal has successfully restored many rivers from Maine to California. Where refurbishment of dams meant new and more efficient generators to produce hydropower, a cost–benefit analysis prevented decommissioning.[48] Such efforts increased hydropower's installed capacity from 56 gigawatt-hours in 1970 to 102 gigawatt-hours in 2017. As a percentage of total U.S. electricity generation for the same years, however, it has fallen from 12 percent of a total of 7 percent. The rapid growth of natural gas power plants and solar and wind technologies have reduced hydropower's share of the nation's electrical-generating capacity.

Notes

1 Atif Ansar, Bent Flyvbjerg, and Alexander Budzier, "Should We Build More Large Dams? The Actual Costs of Hydropower Megaproject Development," *Energy Policy* 69 (2014): 43.
2 Daniel Klingensmith, "*'One Valley and a Thousand' Dams, Nationalism, and Development*" (New Delhi: Oxford University Press, 2007), 1.
3 Sarah Lewin, "Dams over the Decade," *Scientific American* (July 2015): 16.
4 Patrick McCully, *Silenced Rivers: The Ecology and Politics of Large Dams Ecology and Politics of Large Dams* (London: Zed Books Ltd., 1996), 7–8.
5 Brian D. Richter, et al., "Lost in Development's Shadow: The Downstream Human Consequences of Dams," *Water Alternatives* 3 (2010): 14–42.
6 Christopher Sneddon, *Concrete Revolution: Large Dams, Cold War Geopolitics, and the US Bureau of Reclamation* (Chicago: University of Chicago Press, 2015), 5.
7 Geoffrey E. Petts, "Regulation of Large Rivers: Problems and Possibilities for Environmentally- Sound River Development in South America," *Hampden: United Kingdom, Interciencia* 15, no. 6 (November-December 1990): 388.
8 Ibid., 389.
9 Olimar E. Maisonet-Guzman, "Amazon Battle: Dams Conflicts," *ICE Case Studies*, no. 230 (December 2010): 6.
10 The section of the chapter summarizes the treatment of the controversy found in McCully, *Silenced Rivers*, 292–294.
11 Ibid.
12 Ibid.
13 Ibid., 293.
14 Ibid
15 Ibid., 294.
16 Celio Bermann, "Impasses and Controversies of Hydroelectricity," *Estudos Avancados* 21, no. 59 (2007): 139–142.
17 Hailun Huang and Zheng Yan, "Present Situation and Future Prospect of Hydropower in China" *Renewable and Sustainable Energy Reviews* 13 (2009): 1653.

18 Ibid.
19 Ibid., 1654.
20 Sneddon, *Concrete Revolution*, 40.
21 Michael Schaller, *The United States and China in the Twentieth Century*, 2nd ed. (New York: Oxford University Press, 1990), 53.
22 Sneddon, *Concrete Revolution*, 40.
23 Ibid., 41.
24 Charlton Lewis, "China's Great Dam Boom: A Major Assault on Its Rivers," *Yale Environment* 360 (November 4, 2013), 1.
25 Mara Hvistendahl, "China's Three Gorges Dam: An Environmental Catastrophe?" *Scientific American* (March 25, 2008): 13.
26 Mara Hvistendahl, "China's Three Gorges Dam," 5.
27 Ansar, et al. "Should We Build More Large Dams?," 44.
28 Lewis, "China's Great Dam Boom," 4.
29 Mara Hvistendahl, "China's Three Gorges Dam," 7.
30 Pierre Fortin, "The Hydro Industry and Aboriginal People of Canada," *Hydropower and Dams* no. 3 (2001): 50.
31 Much of the material for the historical narrative that follows appears in: Www.hydro-quebec.com/history-electricity-in-quebec/timeline.
32 Sam Howe Verhovek, "Power Struggle," *The New York Times* (January 12, 1992), SM 16.
33 Ibid.
34 Ibid.
35 Ibid., 17.
36 Pierre Fortin, "The Hydro Industry and The Aboriginal People of Canada,"49.
37 Frans H. Koch, "Hydropower-the Politics of Water and Energy: Introduction and Overview," *Energy Policy* 30 (2002): 1208.
38 Will Braun, "Nation of the Dammed," *This Magazine* 45, no. 2 (September-October 2011): 18.
39 Paul Niebrzydowski, "Top Ten Origins: Dams," *Current Events in Historical Perspective* Columbus: Ohio. (2017) at http://origins.osu.edu/connecting-history/top-ten-origins-dams, p. 17.
40 John J. O'Neill, "The Life of Nikola Tesla," *Electrical Engineering* (August 1943): 352.
41 Hydropower Program: The History of Hydropower Development in the United States. www.usbr.gov/power/edu/history.html.
42 Department of Energy, Oak Ridge National Laboratory.
43 David Ekbladh, "Meeting the Challenge from Totalitarianism: The Tennessee Valley Authority as a Global Model for Liberal Development, 1933–1945," *The International History Review* 32, no. 1 (2010): 50.
44 Ibid., 51.
45 Ibid., 4.
46 Sergio Pacca, "Impacts from Decommissioning of Hydroelectric Dams: A Life Cycle Perspective," *Climatic Change* (2007): 282–283.
47 Peter H. Glieck, "*Dam it, Don't Dam It: America's Hydropower Future*," www.huffingtonpost.com/peter-h-glieck/americas-hydropower-future_b_1749182.html.
48 Ibid.

References

Ansar, Atif, Bent Flyvbjerg, and Alexander Budzier. "Should we Build More Dams? The Actual Costs of Hydropower Megaproject Development." *Energy Policy* 69 (2014): 43–56. doi: 10.1016/j.enpol.2013.10.069.

Bermann, Celio. "Impasses and Controversies of Hydroelectricity." *Estudos Avancados* 21, no. 59 (2007): 139–154. http://dx.doi.org/10.1590/S0103-40142007000100011.

Braun, Will. "Nation of the Dammed." *This Magazine* 45, no. 2 (September-October 2011).
Ekbladh, David. "Meeting the Challenge from Totalitarianism: The Tennessee Valley Authority as a Global Model for Liberal Development, 1933–1945." *The International History Review* 32, no. 1 (2010): 47–67. doi: 10.1080/07075330903516637.
Fortin, Pierre. "The Hydro Industry and Aboriginal People of Canada." *Hydropower and Dams* 3 (2001): 47–50.
Glieck, Peter H. "Dam it, Don't Dam It: America's Hydropower Future." *Huffington Post*, December 6, 2017. www.huffingtonpost.com/peter-h-glieck/americas-hydropower-future_b_1749182.html.
Huang, Hailun, and Zheng Yan. "Present Situation and Future Prospect of Hydropower in China." *Renewable and Sustainable Energy Reviews* 13, nos. 6–7 (2009): 1652–1656. doi: 10.1016/j.rser.2008.08.013.
Hvistendahl, Mara. "China's Three Gorges Dam: An Environmental Catastrophe?" *Scientific American*, March 25, 2008. www.scientificamerican.com/article/chinas-three-gorges-dam-disaster/.
Klingensmith, Daniel. *'One Valley and a Thousand': Dams, Nationalism, and Development*. New Delhi: Oxford University Press, 2007.
Koch, Frans H. "Hydropower: The Politics of Water and Energy: Introduction and Overview." *Energy Policy* 30, no. 14 (November 2002): 1207–1213. doi: 10.1016/S0301-4215(02)00081-2.
Lewin, Sarah. "Dams Over the Decade." *Scientific American* 313 (July 2015): 16. doi: 10.1038/scientificamerican0715–16.
Lewis, Charlton. "China's Great Dam Boom: A Major Assault on Its Rivers." *Yale Environment* 360, November 4, 2013. https://e360.yale.edu/features/chinas_great_dam_boom_an_assault_on_its_river_systems.
Maisonet-Guzman, Olimar E. "Amazon Battle: Dams conflicts." *ICE Case Studies* 230 (December 2010). http://mandalaprojects.com/ice/ice-cases/amazondams.htm.
McCully, Patrick. *Silenced Rivers: The Ecology and Politics of Large Dams Ecology and Politics of Large Dams*. London: Zed Books Ltd., 1996.
McElroy, Michael B. *Energy: Perspectives, Problems, & Prospects*. New York: Oxford University Press, 2010.
Niebrzydowski, Paul. "Top Ten Origins: Dams." Accessed November 5, 2018. http://origins.osu.edu/connecting-history/top-ten-origins-dams.
O'Neill, John. *Prodigal Genius: The Life of Nikola Tesla*. Hollywood, CA: Angriff Press, 1981.
Petts, Geoffrey E. "Regulation of Large Rivers: Problems and Possibilities for Environmentally-Sound River Development in South America." *Interciencia* 15, no. 6 (November-December 1990): 388–410.
Richter, Brian D., et al. "Lost in Development's Shadow: The Downstream Human Consequences of Dams." *Water Alternatives* 3, no. 2 (2010): 14–42.
Schaller, Michael. *The United States and China in the Twentieth Century*. New York: Oxford University Press, 1990.
Sneddon, Christopher. *Concrete Revolution: Large Dams, Cold War Geopolitics, and the US Bureau of Reclamation*. Chicago: University of Chicago Press, 2015.
Verhovek, Sam Howe. "Power Struggle." *The New York Times*, January 12, 1992, SM 16.

8 Solar power

Capturing the power of the Sun

Introduction

Solar power became the world's leader in new electricity generation in 2017. Worldwide, public utilities and private companies installed 73 gigawatts (1 gigawatt is 1 billion watts) of new solar photovoltaic capacity. Wind followed with 55 gigawatt-hours of new capacity, coal with 52 gigawatt-hours, natural gas with 37 gigawatt-hours and hydroelectric with 28 gigawatt-hours. However, its contribution to the world's energy supply recently passed the 1 percentage point. Installed capacity remains dominated by fossil fuels with oil at 33 percent, coal at 30 percent, natural gas at 24 percent, hydro at 7 percent, nuclear at 4 percent, and solar and wind at 2 percent.

Technological advances in production resulted in significant declines in the price of solar panels with China becoming the world's primary supplier. Additionally, the Sun, recognized as Earth's source of light and heat now became scientifically acknowledged as the galaxy's vast nuclear fusion reactor. In effect, Earth receives 885 million terawatt-hours each year, or 165,000 terawatts (1 terawatt equals 1 million megawatts) of power from the Sun every moment of every day. For the last 4 billion years and for another many billion more, "we are bathed in energy."[1] U.S. consumers use more energy per capita than any other consumers globally, and its total consumption for 2018 was a bit more than 18 terawatt-hours. So, the truly big resource for humanity's future energy needs exists in the sky in the form of that great big nuclear hydrogen fusion reactor. With a global population projected to reach 10 billion by 2050, harnessing 60 terawatt-hours will provide each individual with the modern equivalent of a few kilowatt-hours (1 kilowatt-hour is 1,000 watts) each day. Operating continuously for a day, a 40-watt electric appliance uses 1 kilowatt-hour. Theoretically, such goals are achievable with state-of-the-art solar panels covering a landmass equivalent to the size of Texas for the entire world.

The early history of solar power

Pointing a sheet of glass with silicon at the Sun (a solar panel) and out the back of the panel, you get electrical energy. In addition, an array of solar panels can be both centralized powering a grid or installed house-to-house making each a power grid. Such flexibility cannot be achieved by either coal-fired or nuclear power plants. In history, capturing the Sun's energy spans many centuries. Using glass and mirrors to concentrate and reflect sunlight set fires and burned torches in ancient civilizations including China, Greece and Rome. The latter's bathhouses, homes and public buildings captured the Sun's warmth with large, southward-facing windows from the first to the fourth centuries CE. Sunrooms became commonplace. The Anasazi people, ancestors to the Pueblo in North America, in 1200 CE lived in cliff dwellings that faced south the capture the winter sun.

Fast-forward to the modern era, discoveries and inventions brought us the solar photovoltaic panel. In 1767, the Swiss scientist Horace de Saussure built the world's first solar collector on an expedition to South Africa to cook food. In 1816, a Scottish minister, Robert Stirling, built heat engines that concentrated the Sun's thermal energy to produce power. A major breaking discovery in 1839 of the photovoltaic effect occurred when the French scientist Edmond Becquerel placed two metal electrodes in a liquid electricity-conducting solution that generated electricity when exposed to light. William Grylls Adams discovered that selenium produced electricity when exposed to light. The photoconductivity of selenium proved that light without heat could be converted to electricity. An

Figure 8.1 The organic energy provided by mules pulling wagon trains of iron-ore in late nineteenth century Ketchum, Idaho contrasts with the energy of the sun captured by modern solar panels.

American inventor, Charles Fritts, used selenium wafers to build the first solar cells. The first solar water heater followed in 1891. Before that invention, however, Heinrich Hertz discovered that ultraviolet light altered voltage and caused a spark to jump between two metal electrodes.

The 1904 World's Fair in St. Louis featured a solar array so powerful that it inadvertently fried birds flying 40 feet aboveground. Between 1905 and 1921, Albert Einstein publishes papers on the photoelectric effect. In 1953, Gerald Pearson, a physicist at Bell Laboratories, discovered that silicon was more efficient than selenium in making a solar cell. His two colleagues, Daryl Chapin and Calvin Fuller, refined Pearson's discovery and developed the silicon photovoltaic (PV) cell that converts enough of the Sun's energy to run daily electrical equipment. *The New York Times* reported that the engineers' discovery was "the beginning of a new era, leading eventually to the realization of harnessing the almost limitless energy of the sun for the uses of civilization."[2]

The costs of production, however, prohibited the entry of the silicon cell into the commercial market. In 1956, a single-watt solar cell cost almost US$300 per watt to build. At the same time, building a commercial power plant cost 50 cents a watt. With commercial and residential applications of solar power prohibitive on a cost basis, toy manufacturers installed miniature solar cells to power toys, such as model DC-4 propellers and ships in wading pools.

While commercial efforts failed, both the U.S. Army and the Air Force believed that the solar cell offered the best solution for powering its top-secret earth-orbiting space satellites. The Pentagon awarded the navy the contract to launch the first satellite, and it regarded solar cells as an untried technology, choosing hydrogen fuel cells instead. Over the objections of Hans Ziegler, the world's foremost authority on satellite instruments, the navy agreed to install two power systems into the launching of Vanguard 1. As a compromise solution, hydrogen fuel cells and silicon photovoltaic cells powered the launch. As Dr. Ziegler predicted, the hydrogen fuel cells failed during orbit while the solar cells kept Vanguard communicating for years. In the United States, the Vanguard I and Explorer II space missions and, in the Soviet Union, the Sputnik-3 satellites were launched with an array of PV cells in 1958–59. In 1964, the National Aeronautics and Space Administration (NASA) used a photovoltaic array to power its first *Nimbus* spacecraft and from that time forward used increasingly powerful solar arrays in space exploration. A year later, the idea of a satellite solar power station resulted in the launching of the first Orbiting Astronomical Observatory powered by a 1-kilowatt array of PV cells. By the late 1960s, solar power provided the energy required to send satellites into space.

Space programs continued to flourish in the 1960s and early 1970s. Financed by the U.S. Congress, the cost of solar cell technology for military purposes was absorbed by taxpayers. It remained too costly for commercial and residential use until Dr. Elliott Berman, with financial support from the Exxon Corporation, designed a less costly solar cell using cheaper materials. By lowering the costs of energy from US$100 a watt to US$10 a watt, electricity for commercial and residential use became available to those living a distance from power grids, for oil rigs located on- and offshore, lighthouses, railroad crossings and a few domestic appliances. The technology that led to the production of cheaper solar cells made energy from the Sun affordable. As the oil and gas industry purchased solar panels in increasing numbers, the incipient solar cell industry received the financial capital to expand and eventually flourish. Continuing reductions in wholesale costs of roughly US$1 a watt made affordable silicon PV solar panels possible. For a completed installation including labor and materials, US$5 a watt became the industry standard.

Within a few years, solar cells became the industry standard and those who believed that they would always be a stopgap measure until nuclear energy became a standard for powering orbiting satellites proved to be wrong. Nuclear energy never powered more than a few satellites. Space applications became the impetus for the development of more power arrays of solar cells. Without them, the world's telecommunications revolution may not have happened, and commercial satellites broadcasting information, bringing news to populations without electrical grids, sending alerts of impending and dangerous weather events and much more, depended on solar energy. The technological revolution of the last half century would have stalled without increasingly powerful arrays of solar panels.

Yet, before solar energy could begin to fulfill its promise as a renewable source, history would intervene in the form of the Arab oil embargo of 1973. It was triggered by the Six-Day War with Israel that alerted the world, including the United States, that dependence on oil from the Middle East made them vulnerable to circumstances beyond their control. During the presidential debates in 1976, President Gerald Ford and Governor Jimmy Carter argued about the best way forward in addressing the oil crisis caused by the embargo. Future president Carter argued,

> We're gonna run out of oil. We now import about 44% of our oil. We need to shift from oil to coal. We need to concentrate on coal burning and extraction, with safer mines, but also clean burning. We need to shift very strongly toward solar energy and have strict conservation measures. And then as a last resort only, use atomic power.[3]

Reading that proposal today reveals the knowledge gap that existed among proponents of coal consumption and its toxic effects on the global climate system. However, Governor Carter, a nuclear engineer, proved to be prescient in recognizing the future role of solar energy.

On March 1, 1977 the now-president Carter (1977–1981) sent to Congress a bill to create a new Department of Energy. He noted,

> It was like pulling teeth to convince the people of America that we had a serious problem in the face of apparently plentiful supplies, or that they should be willing to make some sacrifices or change their habits to meet a challenge which, for the moment, was not evident.[4]

Once passed by Congress, the new federal department with legislative mandates began the process of instituting a series of far-reaching new regulations. Electric utility companies were required to join an effort to better insulate buildings and homes. Higher-efficiency home appliances would be required. Regulations imposed new fuel-efficiency standards for automobiles and the use of pollution-control devices. Synthetic fuel production and carpooling with tax incentives became part of a new energy strategy. He also deregulated energy prices and launched a program to develop synthetic fuels. The deregulation of natural gas prices increased exploration for new supplies and reduced waste of this clean-burning fuel. The new bills also included strong encouragement for solar-power development and tax incentives for the installation of solar units in homes and other buildings.[5]

President Carter lowered the thermostats in the White House, and at other federal buildings, installed solar panels on its roof and a wood-burning stove in the living quarters to get the country's citizens to change their behavior. Undoubtedly, some complained, but as he noted, "it was like pulling teeth." His successor, Ronald Reagan (1981–1989), had the solar panels removed and the thermostats turned up and denied the realization of an impending energy crisis. These symbolic changes represented a setback for the new Department of Energy's (DOE) renewable energy initiative but not its undoing. President Reagan cut the budget of the Solar Energy Research Institute (SERI) by 90 percent, effectively decimating its workforce and setting back its research program. His action effectively stalled the research on PV begun in 1977 during the Carter years by SERI. President George H.W. Bush (1989–1992) reinstated SERI's budget and renamed it the National Renewable Energy Laboratory (NREL).

Among NREL's many centers, the National Center for Photovoltaics (NCPV) is of interest because of its state-of-the-art research. It focuses on boosting solar cell efficiencies and lowering the cost of solar cells, modules, and systems. With scientific advances and industry support of NCPV, scientists and engineers serve as a foundation for valuable collaborations and support from universities and PV users. The larger national Solar Energy Program collaborates with universities and the solar industry to make large-scale solar energy systems cost-competitive with other energy sources by 2020. To achieve this goal by 2020, The DOE launched its SunShot Initiative in 2011, with between US$250 and US$270 million annual support, for NERL's deployment of millions of solar installations operating across the country. Since 2011, the installed price of utility-scale solar power had fallen 70 percent, and the installation of solar modules has grown more than 10-fold in the United States, making solar energy widely affordable and available to more American homes and businesses. *On the Path to SunShot*, a report of the initiative's achievements and future goals, one goal stands out. The total installed cost of solar energy systems was lowered to US$.06 per kilowatt-hour for residential energy. By the end of 2017, solar energy closed in on 2 percent of the nation's electricity-generating capacity, with an achievable goal of 10 percent in a few years.

Global initiatives

By 2017, countries around the world added 75 gigawatt-hours of solar photovoltaic capacity. To put this addition in proper perspective meant that installers made 31,000 solar panels operational every hour. The installed capacity in 2016 exceeded the cumulative output of the five previous years and a 45 percent increase from 2015 to 2016. Growth in Asia accounted for about two-thirds of the world's additional capacity. Specifically, China, Japan, India, the United Kingdom and the United States made up 85 percent of the additions. China dominated the solar PV energy market in manufacturing and use while countries on other continents contributed to the growth in solar power. By 2017, every continent, except Antarctica, had installations providing at least 1 gigawatt-hour, while 24 others exceeded that figure. Germany, Japan, Italy, Belgium and Australia led in providing solar PV capacity to each individual.[6]

China

Given the country's population that exceeds one billion people, on a per capita basis its solar energy profile lags many countries including Germany, Japan, Italy, Belgium and Australia. However, China's determination to add solar PV power to its energy mix remains impressive when compared to other countries. In 2016, it added 34.5 gigawatt-hours, an increase of 126 percent over the previous year. With solar capacity at 77.4 gigawatt-hours, it remained ahead of all other countries. Its rapid growth, however, causes ongoing challenges such as inadequate transmission that it hopes to resolve by building ultra-high-voltage transmission lines connecting northwestern provinces with coastal ones. Alleviating air pollution in many of its industrial cities and reducing carbon dioxide emissions that contribute to climate change highlights the need for a nonpolluting energy source. During the country's modernization, China's awareness sharpened as solar PV's costs became competitive and recognized its potential to satisfy the insatiable need for electricity.

As a result, renewables make up 11 percent of the country's energy use, with a plan to reach 20 percent by 2030. Despite the rapid growth of large-scale solar PV plants, however, they generate only about 1 percent of the country's electrical energy. Like all countries committed to weaning themselves away from fossil energy, a long and difficult journey awaits China in the following decades. In China, experts believe that in 50 years, solar energy will provide 50 percent of the nation's power. Toward that end, the country's National Energy Administration strengthened by the Renewable Energy Law launched its "Forerunner" initiative. It encouraged new solar facilities to use only advanced products with high efficiencies. The initiative sought bids on eight "Forerunner" sites to promote competition on price and lower installation costs. The winning bid guaranteed a per kilowatt-hour of US$0.07, making solar power competitive with coal power.[7]

The challenge has not deterred China from building the world's largest solar PV plant. Begun in 2014 and completed in 2017, the Ningxia solar project in

China's arid northwest autonomous region covers the equivalent of 7,000 U.S. East Coast city blocks. It cost almost US$2.5 billion to manufacture and install 6 million panels. It will generate 2.37 billion kilowatt-hours of electricity annually, equivalent to the power produced by a 400-megawatt-hour coal-fired plant.[8] Other impressive projects point out China's commitment to solar power. A new solar installation in the city of Huainan, in the coal-mining region of Anhui province began operation in May 2017. It generates 40 megawatt-hours of electricity, enough to power 15,000 homes. Interestingly, the installation physically replaces an area that was strip-mined for coal, providing a window into the many ways in which energy flows occur. Land subsidence created a cavity and intense rainfall filled it to create a lake that is between 4 and 10 meters deep (13–33 ft). The solar installation floats on the surface of this lake. The lake's water helps cool the surface of the solar panels, reducing the risk of overheating. Other large-scale projects include using the surface of a fish farm to install 300 solar panels in Zhejiang Province.

The large Longyangxia Dam Solar Park floats on the surface of the reservoir behind the dam in western Qinghai province on the Tibetan Plateau. The combined use of hydropower to generate electricity and a large 27-square-kilometer (10-plus-mi^2) reservoir to float and cool solar panels represents an innovative way to maximize renewable energy resources. It produces 850 megawatt-hours of electricity, enough to supply as many as 200,000 homes.[9] Two factors have propelled China into a position of leadership in the renewable revolution. Because China is the world's largest emitter of fossil coal pollutants, repeated episodes of

Figure 8.2 Solar panels and wind turbines floating on the surface of a reservoir.

toxic smoke and smog in its industrial cities and its capital city of Beijing cried out for action. Enough public anger convinced the country's communist government that decontaminating the atmosphere required an environmental response. China ratified the Paris Agreement, an accord within the United Nation Framework Convention on Climate Change that seeks to limit carbon emissions to 1.5 degrees Celsius (3.6 °F) in this century beginning in 2020. As of September 2018, 175 countries had signed the accord. Unfortunately, the United States withdrew from the agreement before its start date in 2020. A second factor drives China's leadership in renewable energy. Recognizing its potential at home and abroad, China recognizes the economic benefits of spreading its low-carbon technology and products beyond its national boundaries. Andres Hove from the Paulson Institute and based in Beijing believes that the country's leaders saw a "huge investment opportunity" by exporting solar power and electric vehicles to the developing nations in Africa, South Asia and Latin America. China has invested billions in the new global clean energy market in Brazil, Egypt, Indonesia, Pakistan and Vietnam.[10]

Japan

With few coal and petroleum reserves, nuclear power and coal together are projected to supply nearly 50 percent of the country's power by 2030. Its commitment to the 2015 Paris Agreement goals supported a renewable energy target of 22 to 24 percent by 2030 as well. The Great East Japan Earthquake and resulting tsunami on March 11, 2011, that led to the meltdown of the Fukushima nuclear power stations, a major source of the country's electricity, made the case for a commitment to renewables. The Japanese government shut down 50 nuclear power plants across the country for fear of additional disasters. Most reopened after inspection, but anti–nuclear energy critics argued for renewables claiming that a national solar power initiative could generate as much power as 10 nuclear plants. The counterargument prevailed initially suggesting that in a relatively small mountainous country, solar farms were an unrealistic solution. The resulting shutdown led the Japanese government to introduce a series of reforms including the decentralization of the nation's electrical grid.

Its Feed-in Tariff (FIT) program introduced in 2012 targeted renewable energy as a growth area. The tariffs are rebates to utilities with long-term contracts that purchase electricity at predetermined rates from renewable energy producers. The amount the electricity traded on the wholesale market remains relatively low for all energy sources except solar power. In June 2016, only 3.3 percent of all customers had switched from major utilities to new decentralized PV suppliers. By contrast, solar PV on some days and at specific decentralized utilities provided as much as 80 percent of electricity usage. Since the government capped solar FIT regionally, many regions have reached their maximum usage including Hokkaido, Kyushu, Okinawa and Shikoku.

Japan's boom in solar photovoltaics accelerated significantly in 2016 to 5 percent of all energy usage. All renewables together accounted for 14.2 percent of production in that year, with hydroelectric power providing the largest share. Solar and smaller amounts of biomass, wind and geothermal contributed smaller percentages of power. Getting off the central grid in the wake of the Fukushima nuclear crisis has made the difference in solar energy's growth rate. Community-based solar panels and small hydrogen-oxygen fuel cells for heat and power may point the way to a sustainable future in a mountainous flat-land-challenged country. The Fujisawa Sustainable Smart Town relies on community-based renewable energy.

The smart town, located on the outskirts of Tokyo, provides essential services to residents: energy, security, mobility, health care and community. Built for 3,000 residents, the town is zoned for noncar owners and uses an eco-car sharing and rent-a-car service. By not owning a car, the community's living space avoided the need to allocate valuable land for automobiles. With improving air quality, the health benefits for residents became achievable. For heat and electricity, Fujisawa relies on solar panels and small-scale fuel cells owned by the community.

Japan's densely populated and urban nation needs much more renewable energy than can be provided by any number of small smart towns. Its mountainous terrain precludes the installation of numerous solar farms. Like China, floating solar farms on the surface of ponds and reservoirs become one sustainable solution to protect the wellbeing of the environment. As increasing urbanization precludes the further development of large-scale ground-mounted solar farms, using the water's surface to float solar installations becomes a promising alternative. In Japan, strict building codes mitigate the damage caused by earthquakes. The country's prefectures saved financially by installing solar panels on water because they do not require excavation or expensive earthquake-proof foundations. Also, floating farms reduce evaporation and algae blooms in freshwater ponds and reservoirs. Rooftop installations require constant monitoring and energy to avoid overheating, while water serves to cool panels promoting efficiencies. Smart communities located near freshwater, where community members manage their electricity needs, may become buyers of floating solar power.[11] Only Germany and China had more solar installations in 2017 than Japan's 8 gigawatt-hours.

Germany

As the world's fourth-largest economy, Germany has earned the reputation as a solar superpower. Although uncommon among countries making the transition from fossil fuels to renewable energies, the country briefly supplied 90 percent of its power from renewable energy sources for a day in May 2014. This significant event suggests the progress that Germany made only a few years ago. In the next month, June 2014, it achieved another record by supplying more than 50 percent, or 23.1 gigawatt-hours of its energy, with solar power. That amount represented about half of the world's production at the time. Additionally, Germany's 35 gigawatt-hours of solar PV capacity generated 24 gigawatt-hours in one week in May 2014.[12]

Although these impressive figures reflect only a moment in time, they were unique readings and suggested future achievable goals. For a typical year, Germany's level of renewable energy production falls within the 30 percent range, with solar currently providing 7 percent of that amount, followed by wind, biomass and hydropower. With the Fukushima nuclear meltdown in Japan in 2011, Germany began the process of decommissioning its nuclear power plants and canceling plans for additional capacity. Leaving the fossil-nuclear age and promoting PV will become a significant achievement in planning for a future shaped by sustainable power production. In the last 10 years, Germany's growth rate in solar power has nearly tripled.

As it makes the transition, however, Germany's success in producing renewable energy created excess power capacity. Coal-fired and natural gas plants produce most of the country's electricity and they can't be curtailed in response to excess supply on the electrical grid. On windy days with abundant sunshine, too much energy production leads to spikes in excess production. With technological advances in battery storage, unneeded power will result in a decline in greenhouse emissions. In the meantime, too much power in the grid causes the price to enter negative numbers. As a result, the owners of coal and natural gas plants producing excess electricity must pay commercial customers to consume power.[13]

This untenable circumstance led the German Parliament to terminate the subsidy for wind and solar power producers called Feed-in Tariff (FIT). Japan used the program to stimulate research and development in renewables, FIT led to overcapacity. In its place, the parliament substituted an auction system that required utilities to bid on projects to build renewable energy facilities up to a level decided by the government. Rather than a fixed price for power provided by renewables, market-based pricing would become standard practice. In trying to solve the oversupply problem, shutting down polluting power plants became a goal. Yet, coal-fired plants provide power when the wind stops and sunshine disappears. With energy from the grid provided by renewables first, operators make a profit by selling excess power from coal plants to neighboring countries.

Under such circumstances, reducing and eventually eliminating coal-fired plants and curtailing carbon emissions become more difficult. One proposal to rectify the dependence on fossil energy involves the construction of a super grid across the European Union that would promote the use of renewable power across borders. This proposal would eliminate the need for always-on fossil fuel power plants. Although plans exist to construct such a grid, its costs may be prohibitive, ranging from US$112 to US$448 billion depending on the number of participating EU countries.[14]

At the end of 2018, solar PVs accounted for about 7 percent of the country's net electricity generation. During weekdays with ample sunshine, PV can cover 35 percent of the demand while on weekends and holidays coverage can reach 50 percent. Total PV installed power was 2.8 gigawatt-hours distributed over 1.6 million power plants and connected to the decentralized low-voltage grid that is close to consumers keeping transmission costs low. As prices for photovoltaics continue their slide, more than 75 percent since 2006, costs for installations will follow suit. With the average four-person German family consuming about 4,400 kilowatt-hours of electricity a year, advances in solar power technology will meet household needs.[15]

Brazil

As the largest economy in South America, Brazil gets 70 percent of its energy from hydropower. The share of its power from solar energy remains small but the potential for a larger share remains high. Since the country lies between 5 degrees north and 35 degrees south latitude, it is mainly tropical, receiving large amounts of sunshine. With most of Brazil's population living in coastal cities, vast acreage exists for solar farms. Legal and economic barriers prevent solar installation expansion on roofs, office complexes and public buildings. To date, the country has fewer than 20,000 installed solar panels. Bangladesh's landmass is 65 times smaller than Brazil yet uses 1.5 million solar panels to generate electricity.

The absence of government subsidies to support a fledgling industry and policies that deter citizens and entrepreneurs from investing in the solar industry prevent growth and innovation. Although the government supports distributed solar and encourages citizens to generate more power than they need, it taxes them needlessly. It is common practice for those with solar panels to generate 500 kilowatt-hours of electricity and purchase an additional 1,000 kilowatt-hours from the grid. Instead of encouraging distributed solar power by waiving taxes on the 500 kilowatt-hours and taxing the 1,000 kilowatt-hours from the grid, utilities tax the entire amount. Incentives, rather than disincentives, would promote solar power development.

Since 2018, 10 of Brazil's state governments moved to eliminate taxes on distributed solar. Yet government planning and subsidies lag the efforts made by foreign competitors to jump-start solar installations. Of the many metal and thin-film components used to assemble panels and modules, the national government imposes high tariffs on foreign manufacturers. Rather than exempting taxes on new energy sources, state and national governments impose social security taxes and nonspecific taxes that support private- and public-sector workers. Offering a six-year tax break for renewable energy companies and a feed-in tariff approach that other countries offer could put Brazil on a path toward achieving its goal of increasing solar power capacity to 7 gigawatt-hours by 2024.

The country faces the challenge of shifting some of its dependence on hydropower and oil and toward solar power. To meet the challenge, South America's largest solar facility in Pirapora in the southeastern state of Minas Gerais became operational in 2018. Covering 1,200 soccer fields, with a capacity of 400 megawatt-hours, 1,200,000 solar panels will provide power for 420,000 households. "It's a key project of exceptional dimensions at a location that has the advantage of being flat, with little vegetation, a lot of sun, and proximity to a high voltage transmission line," said an executive of the French energy company EDF Energies Nouvelles, which operates the plant.[16] To maximize the plant's ability to capture as much sunshine as possible, the panels sit almost 4 feet (1.2 m) aboveground. They pivot with Earth's rotation, horizontal at midday, tilting as the angles change.

It is a first major step in reaching its Paris Agreement goal of making 45 percent of its energy renewable by 2030. The path to that goal will remain problematic until the government makes policy and passes laws to promote solar energy. Presently, only 0.2 percent of its energy comes from solar. The 15-year delay in promoting renewables means that catching up will become a daunting endeavor. The completion of the Pirapora project using locally built panels, however, suggests a higher level of Brazil's technological development in solar energy. As prices for panels and components continue to decline, the gap between early innovators and later entrants will shrink noticeably. A combination of steep price declines for manufacturing panels and a reduction in taxes offers incentives for entrepreneurs to invest in solar energy.

China, India, Japan and countries in the European Union have begun building solar farms on lakes, reservoirs and other bodies of water. Brazil's long-term commitment to hydroelectric power makes it a likely candidate for using its many reservoirs to become home to solar plants. In 2016, Brazil unveiled the world's first pilot project to produce solar energy from its hydroelectric dams. At Balbina Dam in the state of Amazonas, the floating photovoltaic panels will initially produce as high as 5 megawatt-hours of electricity to power 9,000 homes. Under study is the combined efficiency of solar panels and hydropower and their effect on the water's ecosystem. Once completed, an expected output of 300 megawatt-hours to power 540,000 households seems plausible. As the country's minister of mines and energy has stated, "[w]ith plenty of room to collect solar energy in their reservoirs, our several hydroelectric dams can provide enormous, untapped capacity."[17]

Australia

With massive deserts in the northwest and central parts of the continent, sunshine becomes one of its major resources available to produce affordable energy. The continent receives 58 million petajoules of solar radiation each year, a figure that represents 10,000 times more energy than Australia currently consumes. Despite the availability of this free and renewable resource, solar energy will represent only 1.2 percent of its total energy consumption by 2030. Currently, coal is king in the "land down under." Spending billions of dollars to wean it away from fossil fuels, the national government and its states have invested in large-scale solar power plants.

By using solar thermal and PV technologies, the plan envisions the development of household and commercial capacity of up to 1,000 megawatt-hours of solar power generation. As noted earlier solar PV converts sunlight into electricity directly using PV cells. Its technology makes it adaptable to many uses. Its flexibility ranges from rooftop installations, building designs, vehicles and megawatt power plants. As noted earlier, photovoltaic arrays can concentrate mirrors to generate power for central grids. Solar thermal on the other hand, converts the Sun's radiation into heat. Thermal energy carried by air and water generates electricity by using steam and turbines.[18]

Australian citizens increasingly use residential solar power. With almost 2 million homes using a solar PV system, the country's installed capacity now exceeds more than one solar panel for each person. With solar prices dropping by 40 percent from 2012 to 2018, a future combining solar power with battery storage seems achievable. Price drops make larger household systems more affordable. Those producing 5 kilowatt-hours and 10 kilowatt-hours at no appreciable increase in cost can replace systems that once produced 2 kilowatt-hours. As such, households produce enough energy to limit the amount of power purchased from the grid. By using PV to generate about 30 percent of one's electrical needs, paying for a 5-kilowatt-hours system will take approximately five years or less.[19]

With rooftop installations booming, a total of 111 megawatt-hours of new panels represented a 69 percent rise from February 2017 to 2018. Being one of the top five months in the history of solar installations in Australia, it reflected the significant drop in prices and a renewed effort by the state governments to accelerate commercial developments. With 30 new industrial solar farms scheduled for completion, New South Wales and Queensland lead with 10 and 18 large-scale projects, respectively, that became operational in 2019. Combined these new commercial enterprises will add between 2.5 gigawatt-hours and 3.5 gigawatt-hours to the national grid. Rooftop installations will add another 1.3 gigawatt-hours.

In Queensland, household installations represent the state's largest source of energy. Thirty percent of the homes use energy from solar, the most in the country. In New South Wales, its 10 new solar farms will generate 1.2 gigawatt-hours of electricity and will reduce carbon emissions by 2.5 megatons, the equivalent of removing 800,000 internal combustion vehicles from the roads. The chief executive officer of the Smart Energy Council put Australia's commitment to solar energy in the following way:

> Solar is the cheapest way to generate electricity in the world – full stop. A solar array, at an average size for an average home, if you amortize the cost over twenty years, the effective rate is five cents per kilowatt-hour. That's called an economic no-brainer.[20]

Australians added 1.2 gigawatt-hours of solar PV in 2017. Much of it is in rooftop solar but with additional capacity in large-scale solar farms. Households, community centers, schools and small businesses received incentives by embracing renewable energy and taking control of their electrical bills. The country's Renewable Energy Scheme issues certificates for installed renewable energy systems. The certificates obligate the Scheme to buy electricity from Australians on a quarterly basis. With a 41 percent increase in installed capacity in 2017 compared to 2016 across all of Australia's states and territories, 2018 became a banner year for continued increases, according industry analysts.

The Desertec project

A group of European countries led by Germany considered bringing solar energy from North Africa to meet their goal of reducing carbon emissions into the atmosphere by 25 percent in 2025 when compared to 1990 levels. The ambitious Desertec Industrial initiative in 2013 failed because of political instability across North Africa and a belief among some of its supporters that Europe had enough renewable energy at an affordable cost. So, this grand plan with Germany's support never materialized as the European Union and its investors withdrew support. The Paris Agreement a few years later, in 2015, provided justification for renewed enthusiasm of the original initiative. As explained by London's energy giant Nur Energie, which specializes in solar energy,

> [t]he project is part of the solution to Europe's increasing urgent challenges in the energy sector: meeting the Paris Climate Agreement emission reduction targets, replacing obsolete fossil fuel and nuclear power plants, reducing reliance on imported fossil fuels, and meeting the expected surge in electricity demand for electric vehicles.[21]

Working with the Tunisian government, TuNur, the company created by Nur Energie for the Tunisian project plans to export 4.5 gigawatt-hours of solar power from the northeastern edge of the Sahara to Europe, enough for 5 million homes or more than 7 million electric vehicles. Another factor influenced Nur's decision to build a solar plant in the Sahara. The Malta–Sicily Interconnector, a 120-kilometer (74.5-mi) high-voltage (direct-current) underwater cable connected Tunisia to Europe through Italy. Commissioned in 2015 with a completion date 2020 and after the demise of Desertec, it would become the first of three high-voltage cables providing electricity to Europe via Italy and France to Germany and other participating European Union countries. As noted by TuNur's chief executive, Kevin Sara,

> [t]he economics of the project are compelling: the site in the Sahara receives twice as much solar energy compared to sites in central Europe, thus, for the same investment, we can produce twice as much electricity. We will always be a low cost producer, even when transmission costs are factored in.[22]

TuNur plans to use an energy-producing system called concentrated solar power (CSP) technology, similar to the SunShot Initiative in the United States that can supply solar power on demand coupled with thermal storage. CSP programs seek to make electricity generation competitive with traditional sources with a price of 0.06/kilowatt-hour by 2020. It works by using an array of mirrors to collect sunshine and reflect it onto a central tower. Molten salt stores the energy and releases it on demand. The Malta–Sicily Interconnector plant will generate 250 megawatt-hours of electricity in this way and become one of the world's largest thermal solar facilities. Once the three-cable project is completed, it will be three times larger than Manhattan borough in New York City. Building the solar plant will not only provide energy-starved Europe with renewable power; it will also benefit Tunisia, which currently relies on Algeria during power outages. The plant will also bring electricity to Tunisia's poorer interior.

Originally Morocco was a participant in the now-defunct Decertec project. Now, its goal is to provide electricity to some of Africa's 600 million people who currently rely on wood-burning stoves and diesel generators for heat and light. Since both forms of energy emit carbon dioxide and contribute to climate change, Morocco's Noor 1 solar plant will serve as a clean and renewable form of energy. Located 120 miles from Marrakesh, on the edge of the Sahara, the 160-megawatt-hour CSP plant used 500,000 parabolic mirrors to collect sunlight for the purpose of heating liquid that's creates steam to power turbines. Morocco's goal is to generate 42 percent of its power from wind and solar by 2020.[23]

Conclusion

As global dependence on fossil fuels wanes, technological developments, economies of scale and rapidly declining prices suggest a bright future for solar energy. Globally, solar installations increased 29 percent in 2017 to 98.9 gigawatt-hours up from 76.5 gigawatt-hours the previous year. Accelerated installations currently represent only a fraction, however, of total energy consumption. In 2018, fossil fuels dominated energy use at the global level, accounting for 75 percent of the total. The energy mix for each country differs, however, with the United States using more oil, 37 percent of the total, followed natural gas, with 30 percent, and coal with 15 percent. Nuclear power contributed 9 percent of the total. The remaining 9 percent is provided by hydropower, wind, solar and biomass. By 2020, solar power will represent 3 percent and 5 percent a few years later in 2022. Projecting outward to 2050, hydrocarbons will continue to dominate the energy mix in the United States, providing more than 70 percent of the energy. Presently, China depends on its domestic coal supplies and its coal imports from Australia, which represent 60 percent of its energy. It is projected to decline to 54 percent by 2020. With its global dominance in the manufacture of solar panels, renewable energy, now at 15 percent, will reach 20 percent by 2022. Nuclear energy is negligible at 1 percent.

The French experience reflects its long-standing commitment to nuclear energy in response to the 1973 oil shocks, with usage close to a 50 percent, oil at 32 percent, natural gas at 15 percent and coal at 3 percent, complete the hydrocarbon mix. Wood for household heat and cooking contributes almost 4 percent of the country's energy mix, reflecting the rural composition of a segment of the country's population. Hydropower, wind and solar add up to 3 percent. The energy mix should not be confused with the power generated to produce electricity since it does not account for matters surrounding energy use for transportation, industry and households. Taking this issue into account, France used 74 percent for nuclear, 15 percent for hydropower, renewables 7 percent and fossil fuels, mainly natural gas 10 percent. Each country's experience differs but despite the optimistic rise in solar power, hydrocarbon usage and, for France, nuclear power will continue to dominate the energy landscape.[24]

Combined with wind technology (a topic of the next chapter), they represented 5.5 percent of all electrical generation and, importantly, almost half of all new capacity globally. Both have advantages over the cost of new coal- and gas-fired facilities. As noted earlier, Australia's commitment to renewable energy will reach 100 percent by 2033. By adopting electric vehicles for transportation and electric heat pumps for cooling and heating across the world, a cleaner energy future is possible. "Beyond this, we can develop renewable electric-driven pathways to manufacture hydrocarbon-based fuels and chemicals, primarily through electrolysis of water to obtain hydrogen and carbon capture from the atmosphere, to achieve major emission reductions."[25]

Notes

1 Richard E. Smalley, "Future Global Energy Prosperity: The Terawatt Challenge," *Materials Research Society Bulletin* 30 (June 2005): 415.
2 www.experience.com/advice/careers/ideas/the-history-of-solar-power/.
3 The First Carter-Ford Presidential Debate, September 23, 1976.
4 Jimmy Carter, *Keeping Faith, Memoirs of A President* (Fayetteville, University of Arkansas Press 1985), 94.
5 Ibid., 107.
6 REN21 in 2017, Connecting the Public and Private Sectors on Renewable Energy, "Solar Photovoltaics (PV), 63.
7 He Nuoshu, "Can Brazil Replicate China's Success in Solar," *Chinadialogue* (June 20, 2017): 2.
8 "Out of China's Dusty Northwest Corner, a Solar Behemoth," *Bloomberg News*, September 19, 2016 at www.bloomberg.com/news/articles/2016-09-20/out-of-china-s-dusty-northwest-corner-a-solar-behemoth-arises.
9 *South China Morning Post*, February 21, 2018 at www.scmp.com/news/china/society/article/2096667/china-flips-switch-worlds-biggest-floating-solar-farm.
10 Tom Phillips, "China Builds World's Biggest Solar Farm in Journey to Become Green Superpower,"*TheGuardian*(January19,2017),2atwww.theguardian.com/environment/2017/jan/19/china-builds-worlds-biggest-solar-farm-to-become-green-superpower.
11 The material for the section on Japan was taken from the following sources. "Japan's Solar Boom is Accelerating," *Forbes* (January23, 2017); "Japan Sees Potential in Solar Power," *Japan Today* (April 18, 2017); "Fujisawa Sustainable Smart Town Goes into Full-Scale Operation Near Tokyo," *Panasonic Newroom Global* (November 29, 2014).
12 Sara Thompson, "How Germany Became a Solar Superpower," *SunModo* (August 13, 2015), 1–10.
13 Richard Martin, "Germany Runs Up Against the Limits of Renewables," *MIT Technology Review* (May 24, 2016): 2.
14 Ibid.
15 www.ise.fraunhofer.de/content/dam/ise/en/documents/publications/studies/recent-facts-about-photovoltaics-in-germany.pdfFeb 21, 2018. . . germany.html.
16 Louis Genot, "Huge Solar Plant Aims for a Brighter Brazil Energy Output," *Phys.Org* (November 10, 2017): 1.
17 Bianca Paiva, "Brazil Unveils First Dam-Mounted Floating Solar Plant," *Agenciabrazil* (July 3, 2016): 2.
18 "Solar Energy," *Geoscience Australia* at www.ga.gov.au/scientific-topics/energy/resources/other-renewable-energy-resources/solar-energy.

19 Solar Choice Staff, "Is Home Solar Power Still Worth it in Australia in 2018," *Solar Choice* (January 11, 2018): 1.
20 Naaman Zhou, "Australia's Solar Power Boom Could Almost Double Capacity in a Year, Analysts Say," *The Guardian* (February 11, 2018): 2.
21 Tina Casey, "The Desertec Sahara Solar Dream Didn't Die After All," *Clean Technica* (August 11, 2017): 2.
22 Ibid.
23 Richard Martin, "Morocco's Massive Desert Solar Project Starts UP," *MIT Technology Review* (February 8, 2016): 1.
24 Stephen Eule, "EIA's Annual Energy Outlook 2018-the Ups &Downs," *U.S. Chamber of Commerce* (February 14, 2018), 1–4 at www.planete-energies.com/en/medias/close/about-energy-mix.
25 Andrew Blakers, "Solar is Now the Most Popular Form of New Electricity Generation Worldwide," *The Conversation* (August 2, 2017): 5 at http://the conversation.com/solar-is-now-the-most-popular-form-of-new-electricity-generation-worldwide-81678.

References

Blakers, Andrew. "Solar is Now the Most Popular Form of New Electricity Generation Worldwide." *The Conversation*, August 2, 2017. http://the conversation.com/solar-is-now-the-most-popular-form-of-new-electricity-generation-worldwide-81678.

Carter, Jimmy. *Keeping Faith, Memoirs of A President*. Fayetteville: University of Arkansas Press, 1985.

Casey, Tina. "The Desertec Sahara Solar Dream Didn't Die After All." *Clean Technica*, August 11, 2017. https://cleantechnica.com/2017/08/11/desertec-sahara-solar-dream-didnt-die-baaaack/.

"Fujisawa Sustainable Smart Town Goes into Full-Scale Operation Near Tokyo." *Panasonic Newroom Global*, November 29, 2014.

Genot, Louis. "Huge Solar Plant Aims for a Brighter Brazil Energy Output." *Phys.Org*, November 10, 2017, 1.

Geoscience Australia. "Solar Energy." Accessed November 5, 2018. www.ga.gov.au/scientific-topics/energy/resources/other-renewable-energy-resources/solar-energy.

Goode, B. "Japan Sees Potential in Solar Power." *Japan Today*, April 18, 2017.

Martin, Richard. "Morocco's Massive Desert Solar Project Starts Up." *MIT Technology Review*, February 8, 2016, 1.

Martin, Richard. "Germany Runs Up Against the Limits of Renewables." *MIT Technology Review*, May 24, 2016, 2.

Nuoshu, He. "Can Brazil replicate China's Success in Solar." *China Dialogue*, June 20, 2016, 2.

"Out of China's Dusty Northwest Corner, a Solar Behemoth." *Bloomberg News*, September 19, 2016. www.bloomberg.com/news/articles/2016-09-20/out-of-china-s-dusty-north west-corner-a-solar-behemoth-arises.

Paiva, Bianca. "Brazil Unveils First Dam-Mounted Floating Solar Plant." *Agencia Brazil*, July 3, 2016, 2.

Phillips, Tom. "China Builds World's Biggest Solar Farm in Journey to Become Green Superpower." *The Guardian*, January 19, 2017. www.the guardian.com/environment/2017/jan/19/china-builds-worlds-biggest-solar-farm-to-become-green-superpower.

Smalley, Richard E. "Future Global Energy Prosperity: The Terawatt Challenge." *Materials Research Society Bulletin* 30 (June 2005): 412–417.

Solar Choice. "Is Home Solar Power Still Worth it in Australia in 2018." Accessed November 5, 2018. www.solarchoice.net.au/blog/solar-power-for-homes-australia-worth-it-2018.

Thompson, Sara. "How Germany Became a Solar Superpower." *SunModo*, August 13, 2015, 1–10.
Zheng, Sarah. "China Flips the Switch on World's Biggest Floating Solar Farm." *South China Morning Post*, February 21, 2018. www.scmp.com/news/china/society/article/2096667/china-flips-switch-worlds-biggest-floating-solar-farm.
Zhou, Naaman. "Australia's Solar Power Boom Could Almost Double Capacity in a Year, Analysts Say." *The Guardian*, February 11, 2018, 2.

9 Capturing the power of wind

Introduction

Sources of energy change with fluctuating needs, with availability, costs and the density to produce power efficiently. The power of the wind, however, has remained a constant throughout the history of the planet. It made placid seas churn with violent waves. Its force swept across the land, rearranging the topography, creating dunes in some places and eroding hilltops in others. Creation and erosion signaled its power and presence. Dust storms became a signature event by moving topsoil across continents.

In human history, its energy powered filled sails and the blades of early windmills to raise groundwater to the surface and grind grain into flour. Over time the increasing size and efficiency of windmills assumed many more functions. The invention of turbines, a mechanical breakthrough, set the stage for the modern technology of wind power. It becomes the primary focus of this chapter. First, we step back in history to examine its premodern uses. It expands on the abbreviated history of wind power found in Chapter 2, which focused primarily on waterpower.

Wind power for ancient navigation

The energy from oarsmen powered the speed of ancient ships before the invention of the lateen triangular sail. The number of crewmembers rowing determined the speed of the ship. If ancient oarsmen were aided by the wind, the presence of the square sail permitted sailing ahead of the wind. The lateen sail invented in either ancient Egypt or Persia changed that with the fore-and-aft sail, the lateen. Its design allowed it to take the energy of the wind from either side and to tack to the wind to power ships of increasing size. Those who invented it changed the potential impact of the sailing ship on long-distance navigation. Sailing boats appear in the early records of the Eastern Han dynasty in China (25–220 CE).

Ancient wind power for sailing craft

Millennia before capturing the kinetic energy of the wind to power sailing craft became the subject of written accounts and treatises, Polynesians in doubled hulled canoes with sails explored and settled the Pacific Ocean islands and greater Australia.

Before that innovation became commonplace, Egyptian sailors traveled the length of the Nile River 5,000 years ago in boats equipped with sails made of linen and papyrus. Traveling the Nile and transporting people and products from one end of Egypt to another bound the region together. Roman sailing craft equipped with lateen canvas transported grains, wine and olive oil around the empire. The economic benefits of sea and river travel compared to overland travel were apparent to all invested in transporting people and products. Roman records note that the Italian port of Ostia received approximately 1,200 large merchant ships each year, or about five per navigable day. As noted earlier, by pivoting on a ship's mast, the lateen sail allowed ships to sail into the wind. The wind could be gentle breezes or strong winds depending on the direction and planned course of a ship. Increasing the number of masts over time provided maximum propulsion for ships of increasing size. That transition took centuries to develop as the energy of the wind translated into an efficient and effective source of power for ships on the high seas and oceans.

By 1400 CE, shipbuilders mastered the transition by increasing the number of masts on a ship. From then until the nineteenth century, nations depended on sailing vessels that became the clipper ships of maritime lore. Nations conducted commerce and conquest from 1400 CE to 1850 CE using the energy of the wind to power its vessels of increasing size. With fossil coal providing the energy to turn water into steam that, in turn, powered turbines and propelled ships forward, the importance of sailing ships faded quickly. Recreational use became fashionable in developed countries while developing nations continued to use the kinetic energy of the wind for fishing and other commercial activities.

Ancient wind power for terrestrial use

Vertical-axis windmills disappeared mostly with the invention of the now-more-common horizontal-axis version. Yet, despite its ancient origins and its more than 1,000-year history, the Nashtifan windmills continue to operate in modern Iran (ancient Persia). Originally designed to grind wheat into flour, these windmills continue to perform their ancient function. Located on the windy and dry plains of northeastern Iran about 30 miles from Afghanistan, the small town of Nashtifan is home to some of these earliest windmills. A 65-foot-high earthen wall (about 20 m) protects town residents from cutting gale-force winds. Atop the wall, ancient Persians built 24 vertical-axis windmills made of handcrafted wooden blades, clay and straw for milling grain into flour. They perform that function to this very day. Given the region's powerful winds, the blades turn the energy from this force of nature into the power required to turn grindstones. Maintenance and repairs to its turbines remain keys to their continued performance.

Unlike horizontal-axis windmills, those at Nashtifan funnel the wind down the mast to the grindstone without the need for additional gearing. Unlike the horizontal-axis windmills that lift the energy of the wind upward to power a grindstone, the Persian model funnels the air down the mast to the grindstone. "The tall walls framing the windmills both support the turbines and funnel the airflow like an elliptical throat in a primitive wind tunnel."[1]

For centuries in the ancient world, these vertical sails of reed bundles were employed in China, India and other parts of the world by farmers to pump groundwater, grind grain, and crush sugarcane. In twelfth-century China, records provide details of windmills lifting groundwater in the Song dynasty. The Chinese statesman Yehlu Chhu-Tshai documented the use of windmills in 1219 CE.[2] By the seventeenth century, drawings of windmills appear in important agricultural technical books. Its windmills were of the two types described earlier, the vertical-axle and horizontal-axle versions, used mostly in coastal areas. Their size, 7 to 10 meters (23–33 ft) made them similar to those in use at the same time in countries around the world. Scholars recognize independent invention rather than a diffusion of technology as the proper explanation for the existence of windmills.

Although their use continued, the traditional European windmill invented independently consisted of a large vertical post with blades, an axle and a mill affixed horizontally to the post. Called a post mill by inventors, users could swing the entire windmill around the post to capture the energy of the wind. William of Almoner built the first windmill using this design in Leicester, England in 1137 CE. Some commentators regarded the post mill as a social leveler removing control of power sources, mostly waterpower from the hands of royals and land

Figure 9.1 Mill on Wimbledon Common by George Cooke (1781–1834).

Source: Courtesy of the Museum of Wimbledon.

barons. Herbert of Bury, a middle-class businessman from Suffork, stated in 1180 CE, "The free benefit of the wind ought not be denied to any man."[3]

Windmills in the Dutch republic

Continental European countries, including France, Spain, Belgium, Denmark, the Italian states and German principalities, adopted this relatively simple post-mill design. As many as 30,000 windmills produced the equivalent of 1 billion kilowatt-hours of electricity in these European countries in later years of the thirteenth century.[4] On the other hand, the Netherlands, with its topographical challenges of draining land at sea level or below, required a more powerful windmill. As early as 1390 CE, the more sophisticated and powerful tower mill that existed earlier along the Mediterranean Sea replaced the post mill there. As its name suggests, the tower mill was taller than its predecessor and constructed of wood, brick and stone.

The Dutch placed the common post mill at the top of a four-story tower. Stories were devoted to grinding grain, removing chaff, storing the grain and, at the bottom level, providing living quarters for the miller and his family. Forecasting its successor, the modern wind turbine, the tower remained fixed and immovable. A cap at the top contained a rotor, horizontal axle and gearbox. To catch the direction of the wind, an operator manually pushed a large lever at the back of the mill, pointing the rotor toward the wind. Improved sails generated aerodynamic lift and increased rotor speed that allowed for faster and more efficient grinding and pumping. During storms, millers furled the rotor sails to prevent damage to them. Pointing the sails into the wind and folding them in a storm were among the wind master's primary jobs.[5]

Given the height and durability of Netherlands' tower mills, owners and operators expanded their use. Beyond land drainage and the milling of grain, tower mills refined pepper, other spices and cocoa. The wind power of tower mills aided the process of making dyes and paint pigments. They provided power to sawmills and reduced wood pulp to paper. One estimate suggests that they provided 25 percent of Europe's power from the fourteenth to the eighteenth centuries. Waterpower, human and animal labor provided 75 percent. The invention of the steam engine and the discovery and use of cheap coal transformed the energy mix from mostly organic forms of energy to a mineral energy regime.[6] However, 10,000 tower mills dotted the Dutch landscape during its zenith spanning the seventeenth century.

Mobile motion energy of wind became its means for transporting Dutch wares by ship to its trading partners. Its fishing boats benefited greatly from wind energy. Peat, imported fuel and wind energy resulted in very high labor productivity when compared to other nations during and after the medieval period.

Stationary wind motion energy generated by windmills became a symbol of Dutch engineering vitality. It became a real asset in dealing with an unpredictable Baltic Sea and the environmental damage done by digging peat. Digging left large land areas susceptible to saltwater incursions. In a country with many natural inland lakes, digging peat created many more. By the sixteenth and seventeen centuries, draining lakes using the kinetic energy of windmills became commonplace.

A century later windmills not only drained former peat bogs; they also added land to the country's agricultural stock:

> A region as large as Holland north of Amsterdam-Haarlem gained more than 36,000 ha (hectares) a hectare is 2.47 acres of new land between 1540 and 1650 (90% between 1590 and 1640). This increased the existing 108,000 ha old land of this extended region by one third.[7]

The primacy of peat for energy in the Dutch Republic's economic boom during its golden age remains controversial because nonrenewable peat made the Republic dependent on imported firewood and coal for fuel. These energy sources supported its growing residential population and its expanding commercial and industrial capacity. During this same period, the kinetic energy of windmills played an important role in expanding the power of the Dutch economy in the Western world, Southwest Asia and the Pacific Rim.

Industrial uses of windmills played a significant role in the Dutch economy. Every walled Dutch city possessed at least 10, and when Amsterdam found itself without available space for additional windmills,

> the region along the river Zaan (a distance of 15–20 kilometres from Amsterdam) was used for the extra mills and rapidly developed into the area with the most intensive use of stationary motion energy in the pre-industrial world. At its apex, around 1730, the Zann region had nearly 600 industrial windmills in operation.[8]

Windmills continued to flourish in the Dutch Republic into the eighteenth century, draining flooded fields, grinding grain and corn for flour and seeds to make oil. They also powered sawmills to cut lumber for the shipbuilding industry.

Holland's vertical windmills and those located throughout Europe differed from the horizontal ones invented in ancient Persia about 1000 CE. Although a connection between these horizontal and vertical windmills remains unknown, technological knowledge gained by trade suggests a possible link connecting the development of one and its transfer to the other. A linkage between water- and wind-power generation is suggested by "the direct adaptation of right-hand gearing from the vertical-wheeled watermill to the vertical windmill."[9] Significant independent inventions, including the structure of the windmill, the rotational function of the entire mill house, the design of the windmill's sails to catch the wind constituted a "genuine technological revolution."[10] All things being equal, however, windmills never became as profitable as watermills in grinding grain into flour because their energy-producing power was dependent on variable wind speed, their capital costs and maintenance.

In areas that lacked suitable waterways but met economic, social and environmental considerations, windmills became an alternative source of mechanical energy. Economically, windmills needed a population large enough to finance their construction, maintenance and operation. Profitability in grinding grain became a requirement. Socially, the mill required an affluent class of craftsmen to tinker and innovate to improve the mill's functioning and output of flour. Environmentally, the location on a hill or in an area with an unobstructed view to catch the wind remained a basic condition, without which all other conditions would be meaningless.

Wind- and watermills

In Europe, the earliest documented vertical windmill appeared in England about 1190 CE, although competing theories suggest that they also existed in Eastern Europe at this time. In England, the Domesday Book recorded the existence of 6,082 watermills between 1080 and 1086 CE. At the same time, windmills replaced watermills on weaker flowing waterways. To counter this development, investors built additional watermills on major rivers to increase their energy capacity. However, watermills lost to windmill construction on lesser rivers that were used to transport people and product. Their coexistence reflected the ingenuity of entrepreneurs to maximize water and wind to power their economies. In England, they powered sawmills and grinders for turning wood into dye. A windmill powered a paper mill in Birmingham, England, in 1759. Wind power drained the saturated land in English collieries (coal mines). Such mines depended on housing as many as 50 to 100 horses used to power the pulleys and wooden buckets for draining coal mines at the time.[11]

With each incremental improvement in the windmill's sail over a 500-year period, the mill possessed the features of modern wind turbine blades. For the preindustrial world, wind and water provided energy to power local economies:

> Until the early nineteenth century windmills in common use were roughly as powerful as their contemporary water-driven mills. We have no reliable estimates for earlier centuries, but after 1700 European post mills rated mostly between 1.5 and 6 kilowatts, and tower mills between 5 and 10 kilowatts in terms of useful power.[12]

Replacement of windmills began with the adoption of steam engines well into the nineteenth century.

Wind energy in the new world

Although the seal of modern-day New York City contains an image of a windmill, the size of European and Asian windmills failed to adapt to New World conditions. Settlers regarded the tower mills as too large and expensive to maintain. East Coast settlements, on a well-watered terrain, preferred watermills as sources of power for milling grain, sawing lumber and fulfilling former labor-intensive activities with newer forms of organic energy. Two hundred years after the original settlements in North America, more than 50,000 waterwheels supplemented human and animal labor.

As U.S. farmers and ranchers moved westward to plow over virgin land and feed growing herds of cattle, respectively, mechanics invented a new generation of windmills to bring groundwater to the surface. Daniel Halladay began building windmills at his Connecticut machine shop in 1854. Successful sales caused him to move closer to the market in Illinois. Even though competition from more than 1,000 small and large factories, including Aermotor and Dempster, carved up the growing market for windmills, the demand for pumping water exceeded all expectations. The construction of railroads to bind the western territories to the eastern states accelerated the demand for water. Their steam-driven engines required water for efficient operation.

Farmers, ranchers and railroad owners and operators installed more than 6 million mechanical windmills in the United States between 1850 and 1970. For farmers and ranchers, most of the 6 million were small one horsepower wind machines used for household needs and stock watering for cattle and horses. Much larger windmills, with rotors up to 18 meters (59 ft) in diameter, pumped water for railroad engines. The first produced by the many companies were fan-type windmills that used four paddle-like wooden blades. The next fan-type innovation used thin wooden slats nailed to wooden rims. Aermotor and Dempster introduced light steel blades in 1870 that worked so fast that they required gearing to slow them to accommodate standard reciprocal pumps.[13]

Figure 9.2 An early mechanical windmill used by farmers and ranchers to pump water and generate electricity.

Wind energy and electrical power

With so many windmills rising above the relatively horizontal landscape of the American West from the Mississippi River to the Rocky Mountains, their presence became an iconic feature of the region. As windmills began dotting the western landscape, the telegraph in 1844 became the first practical use of electricity. As noted in earlier chapters, once coal-fired plants began generating electricity in cities during the last decades of the nineteenth century its availability brightened households, factories and darkened streets by 1900. "It became an essential element of the urban lifestyle. In this atmosphere of discovery, inventors contemplated the coupling of wind power and electricity."[14]

The brush turbine

The first person to couple the two was a scientist, Charles F. Brush, who made his fortune in electric arc lamps. In the competition to bring light to darkened city streets, electric arc lamps provided bright light on the major thoroughfares and in the downtown of major cities. Tungsten incandescent lamps brightened residential streets and cheap gasoline streetlamps brought light to poorer neighborhoods, alleys and parks.[15] Lighting the dark improved in these many ways, as did the distribution of electricity by networked natural gas pipelines. Steam turbines turned the kinetic energy of water into sources of power. Brush's success in arc lamp lighting encouraged him to harness the power of the wind by building a large wind turbine to generate electricity.

Behind the large Brush estate on Euclid Avenue in Cleveland, Ohio, in 1888, an immense 56-foot-diameter wheel was mounted atop a 40-ton, 60-foot high tower. From the tower turned a 56-foot-diameter rotor. It was composed of a turbine consisting of a weighted lever, driving belts and pulleys of the proper tension and 144 blades made of cedarwood. When directed at the wind, the speed of the wheel reached 500 revolutions per minute, generating 12 kilowatts of power. The current passed through steel conductors to the basement of the Brush mansion that contained 12 batteries with 36 cells each. The 12 charged and discharged in parallel, with each cell containing a capacity of 100 ampere-hours.

The mansion was furnished with 350 incandescent lamps, with 100 of these used daily. Two arc lamps and three electric generators supplemented the home's primary lighting appliances. Without a major maintenance issue after 15 years of operation, Brush's ability to harness the power of the wind to bring light to his large home proved successful. Although the power of the wind comes without cost, the apparatus built under the supervision of Mr. Brush made it cost-prohibitive for general use.[16] After 1900, Brush used his wind turbine infrequently since Cleveland, Ohio, began providing networked electricity. After 1908, Brush stopped using his wind machine.

The wind charger industry

Given the success, complexity and cost of Brush's machine, it was never mass-produced. Developing a small, less expensive windmill became the focus of manufacturers invested in providing a wind machine to homesteaders who flocked to the western territories to farm and raise cattle for eastern consumers. Charles B. Dempster established the Dempster Mill Manufacturing Company in Beatrice, Nebraska in 1878 to build and market one horsepower mechanical wind machines for submersible water well pumps. He sold the essential equipment needed to succeed as homesteaders, hand and piston pumps, towers, flywheel-type gasoline engines, water storage tanks and much more. Although the company was eventually sold to private investors almost 100 years later in 1985, its role in opening the west remained an important part of its legacy.

With so few farm families having access to electricity, other small manufacturers began producing windmills. Gasoline-powered generators called Delco "light plants" provided electricity to a few farmers. The Jacob brothers, Joe and Marcellus, living on their parents' ranch in Montana, converted a windmill designed to pump water to one generating electricity. They produced rotor blades, its feathering system and a powerful generator for their wind turbine. Recognizing the value of their invention, they moved to Minneapolis, Minnesota in 1927 and began producing their reliable and inexpensive machine. During the 30 years that followed, they sold about 30,000 small wind turbines. Their reliability became recognized across the globe and could be found generating electricity as far away as Ethiopia and Antarctica.

Farmers identified the Jacobs' wind turbine as a top-of-the-line product at a time when a cheaper machine would satisfy their needs. Wincharger, another small company, began producing this small wind turbine capable of generating 6 to 110 volts of electricity. It met the needs of farmers to pump water from a well and charge the batteries of their radios to get the news from distant places. By 1945, the company had sold approximately 400,000 machines worldwide. Six other companies produced small turbines capable of powering a radio, a few 40-watt light bulbs and little else. In 1946, the famous Sears, Roebuck and Co. catalog, which sold everything from prefabricated material to build a small house to clothing for all members of the family, began selling a wind turbine called the Silvertone Aircharger for US$32.50.[17]

With only 10 percent of American farm families receiving electricity from the grid, thousands of small wind turbines provided electricity across the American Great Plains. Across the world, small wind turbines provide service to people living in remote outposts. As many as 100,000 small wind turbines provide electricity to nomadic herdsmen in northwestern China.[18]

The Rural Electrification Act, 1936

The passage of the Rural Electrification Act (REA) in 1936 as part of the New Deal legislation passed by the Congress during the Great Depression signaled the end of the wind charger industry. Private utilities avoided investing in rural electrification. It required the construction of a network of wooden poles, stringing wires and installing transformers. Proximity to urban areas encouraged utilities to serve rural communities, but they refused to reach farther into the region's hinterlands to provide electrical service. The passage of REA changed the mix of individual and centralized power. Its passage satisfied both engineers and operators of centralized power. Centralized power operated on the principle of alternating current while wind chargers produced direct current only. This factor placed the latter at a disadvantage when opportunities for connecting the two delivery systems presented themselves. By 1957, all remaining wind electric enterprises lost their place in the market for electricity and went out of business.

The combination of government subsidies provided in the provisions of the REA and the investments by private utilities filled the power gap faced by rural communities. The act promoted the establishment of cooperatives by local farmers to make federally financed loans to bring electricity to farms within their designated region. All efforts failed to convince federal authorities to include wind charger industries in the legislation.

Coslett Palmer Putnam's great turbine experiment

Converting the direct current of wind chargers to the alternating current of centralized power stations, challenged the skill of a young geologist trained at Massachusetts Institute of Technology (MIT) with no experience with wind power. His interest in this problem peaked when he built a home on Cape Cod, Massachusetts. There, he found both high winds and inflated prices for centralized power. His thwarted plan involved building a windmill to power his home and to sell the surplus to a central power plant. Failure ensued when the direct/alternating current issue remained unsolved. Undeterred, Palmer Putnam approached engineers at S. Morgan Smith Company, a manufacturer of large hydroelectric turbines to construct a massive windmill at the crest of Grandpa's Knob near Rutland, Vermont.

First, the Knob had to be stripped of its vegetation, providing a clear path for workers struggling to move 500 tons (90,700 kg) of material to the summit to build the tower, control room and the wind turbine. Two steel blades, each weighing 7.5 tons (6,800 kg) and each 70 feet (21.3 m) long, needed to be carried to the summit and affixed to the tower where the controls and the turbine resided. The entire enterprise became operational when the blades began to turn on October 19, 1941. As *Time* reported,

> Vermont's mountain winds were harnessed last week to generate electricity for its homes and factories. Slowly, like the movements of an awakening giant, two stainless-steel vanes – the size and shape of a bomber's wings – began to rotate on their 100-ft. tower atop bleak Grandpa's Knob (2,000 ft.) near Rutland. Soon the 75-ton rotating unit will begin generating 1,350 horsepower or 1,000 kilowatts – enough electricity to light 2,000 homes.[19]

Unfortunately, one of the turbine's main bearings failed on February 20, 1943, stopping the power sent to Central Vermont Public Service Corporation's electrical grid. Given the nation's military commitments in World War II, a replacement bearing took two years to be manufactured and delivered to Grandpa's

Knob. Not until 1945 did the Smith–Putnam windmill begin again to provide power to Vermont customers. Catastrophe struck on March 26, 1945, when one of the turbine's massive blades broke off and sailed 750 feet to the ground. The brief history of Putnam's effort to provide a sustainable form of electrical power failed. However, it opened a frontier that, in the long run, would not close. As MIT's dean of engineering, Vannevar Bush, noted, "the project conceived and carried through free enterprisers who were willing to accept the risks involved in exploring the frontiers of knowledge, in the hopes of financial gain."[20]

Financial gain remained elusive, but Putnam's experiment represented a milestone in wind turbine development. Its two-blade 1,250-kilowatt wind turbine underestimated the technical challenges of very large complicated wind turbines. However, his prototype became the model for future wind power developments and installations in the United States in the following decades. As one environmental historian pointed out,

> [i]t is hard to understand how engineers, scientists, and politicians, more or less starting from scratch, could seriously believe in the feasibility of giant turbines with tower heights, rotor diameters, and weights exceeding the lengths, wingspan, and weight of a jumbo jet.[21]

In the decades that followed, efforts to make wind power a part of the country's energy mix met with public opposition, a lack of government subsidies and onerous regulations. Complaints ranged from the unappealing aesthetic quality of large turbines close to population centers, noise and increased bird mortality. Regulations included guidelines specifying height, color and design. Nothing, however, could hide large wind turbines from public space and make them affordable to operate. With pubic opposition and the absence of financial subsidies, wind power development stalled into the 1990s. Elsewhere in the world, wind turbine installations received public support and governmental subsidies.

Wind power on the Great Plains and in the Midwest

European initiatives piqued the interest of U.S. entrepreneurs. In addition, state governments in Texas, California and a number in the Rocky Mountains took a leading role in promoting wind power. Turbines became more reliable, more efficient and designed to become less lethal to birds. Utility companies receive financial credits for selling excess wind energy to other providers. This incentive stimulated others to enter the field. Texas and others took advantage of this system and spearheaded renewed interest in wind energy.

The wind energy potential of the Great Plains became apparent as scientists and engineers explored the velocity of Arctic wind rushing southward to the warm waters of the Gulf of Mexico. This natural condition received ample support from farmers in the region who recognized the financial rewards by capturing the wind for generating electricity. While agriculture remained the engine for economic growth in the Great Plains, small farmers became victims of consolidation and the development of agrobusinesses. Formerly, nineteenth-century farmers used thousands of windmills to raise groundwater for irrigation and to generate household electricity. In some cases, electricity illuminated farm households with a few light bulbs.

By using a small amount of land for wind turbines, small farmers generated income by selling excess electricity to power providers. "Ironically, the very area of the country that several decades ago was home to the greatest concentration of windmills in the world will possibly become so again, this time having them pump out electrons instead of water."[22]

The Great Plains and the Midwest continue to lead the country in onshore wind-power expansion. Texas remains the national leader in wind energy, having installed 10,000 megawatt-hours since 2006. It would be ranked sixth in the world if it were a country. Its Roscoe Wind Farm. with an installed capacity of 781.5 megawatt-hours and 627 wind turbines, is the world's largest such facility.[23] When combined with Texas, the Midwest/Great Plains region captured 89 percent of all investments in wind energy in 2017. Iowa, South and North Dakota, Kansas and Oklahoma were major contributors to the total.[24]

Offshore wind projects complement the robust onshore developments. Currently, five offshore projects plan to deliver 490 megawatt-hours of electricity. The Danish company MHI Vestas invested US$35 million in Clemson University's wind turbine project. It plans to build one of the world's largest offshore turbine's that will generate 9.5 megawatt-hours. In 2016, the Block Island Wind Farm in New Shoreham, Rhode Island, composed of five turbines, began delivering lower electricity rates to the island's residents and large number of warm-weather tourists. Atlantic Coast states, including Maryland, Massachusetts, New Jersey and New York, lead the country in developing advanced plans for offshore wind power. The U.S. Department of the Interior, which regulates offshore projects, needs to streamline the approval process that makes way for plans to become projects.

European wind power developments

Denmark

The availability of cheap oil from domestic suppliers and imports from the Middle East as well as abundant coal deposits in the United States curtailed the development of wind turbine technology. Much of this post–World War II history would be blunted by the oil crisis of the 1970s. In postwar Europe, with fewer mineral fuel resources, experiments in wind power flourished while they waned in the United States. With little fossil fuel and abundant wind, Denmark became a leader in wind turbine technology. Building upon its long history as seafarers, Danes used the power of the wind to explore the North Atlantic and carry its sailors to North America. By 1890, about 3,000 windmills "provided the equivalent of half again as much energy as all the animal power then supporting Danish agriculture."[25]

Future technological developments built on the pioneering work of Poul La Cour, the country's leading advocate of wind power. His early twentieth-century design of a simple, small, and powerful windmill called a *Klapsejler* (clap-sailor) introduced electricity to the country's rural agricultural population. La Cour's design produced direct-current power. By 1906, his 40 wind turbines provided electricity for the country's expanding centralized grid. By 1918, one-fourth of all rural power was provided by wind turbines. The wind turbines looked like traditional windmills except that they had five wide blades covered with metal shutters rather than cloth sails. Using the recent research on aerodynamics, opening and closing the shutters controlled the wind turbines.[26]

A Danish technician, Johannes Juul, a student of la Cour's, began researching wind power early in his career. In 1949, he built a 15-kilowatt wind turbine producing alternating current and connected it to the Danish grid. By 1956, Juul had built a 200-kilowatt wind turbine with a rotor 24 meters (79 feet) in diameter. Through years of experimentation, Juul's design became the prototype for others to follow. It consisted of three fixed, rigid blades supported by several rods. Its rotor maintained an upwind position operating at a medium speed. Low-cost, simple mechanics and safety described Juul's design.[27]

Gradually, Denmark moved from individual solitary wind turbine installations to land-based spatial installations in the decades to follow. By the 1980s and working in harmony with the surrounding landscape, designers viewed wind farms as "a gigantic sculptural element in the landscape, a land-art project."[28] On the flat and open landscape that identifies much of Denmark, engineers and builders working in consort with designers and architects erected wind farms in either rows or in modular design. Coherent clusters rather than scattered single wind turbines became the norm.

The land placed limits on the size and geometry of wind farms. Offshore wind farms in Denmark faced fewer constraints. As such, the country's Ministry of the Environment and Energy embarked on an ambitious plan to build five large offshore wind farms during the first decade of the twenty-first century. This development followed a pattern of capturing the wind from the North Sea that began as early as 1991 with the placement of the world's first offshore wind farm with eleven turbines.[29] When completed, the ministry's plan of 3,500 2-megawatt-hour turbines each produced a combined 15 to 18 billion kilowatt-hours of electricity, or about 50 percent of Denmark's consumption.

With 6,000 land-based wind turbines functioning in the first decade of the new century, Denmark's ambitious initiatives continued to depend on ownership by cooperative associations. Many individuals also own land-based wind turbines. About 100,000 households, or 5 percent of the population, own shares in wind farm cooperatives. However, offshore wind farms with their much larger array of turbines became the domain of power companies. Encouraged with tax incentives from the government and with more robust wind resources, they faced little competition from private investors. However, the country's banks provided loans for much of the installed costs of wind projects and became known as "wind banks."

Land-use planning has made wind turbines ubiquitous in Denmark. They exist almost everywhere in urban and rural settings, on lakes and offshore. Individual wind turbines and small clusters dot the land, city and seascape. To protect scenic areas, a government agency estimated that this small North Sea nation could comfortably absorb between 1,000 to 2,800 megawatt-hours of wind-generated electricity. Some estimates placed them as high as 6,500 units generating 15 percent of the nation's total energy consumption.

As older, smaller and unreliable turbines are replaced with larger and more efficient ones, the government's Energy Action Plan expects 50 percent of the country's electricity to be generated by wind. At current rates, Denmark's targeted goal of 50 percent will be met by 2020. On February 22, 2015, Denmark generated 97 gigawatt-hours from wind, 70 gigawatt-hours coming from onshore wind turbines and 27 gigawatt-hours from offshore installations. On a very windy day on July 15, 2015, its wind farms produced between 116 and 140 percent of the nation's electricity requirements. Eighty percent of the excess was shared with Germany and Norway, with Sweden getting 20 percent. The preceding occurred on a particularly windy day. If sustained winds reached 93 kilometers per hour regularly, the country would have no energy requirements from other sources.[30]

Germany

Historically, domestic coal supplies provided Germany with its electricity. Self-sufficiency in coal with the world's seventh-largest coal reserves and the world's third-largest coal producer supplied Germany with 75 to 90 percent of its energy needs. It continues to be the world's largest producer of lignite, soft brown coal, for the international market. In the 1960s, the coal industry employed 600,000 miners; four decades later, fewer than 70,000.[31] The country's major political party, the Social Democratic Party (SDP) provided subsidies to the coal industry on economic and national security grounds. Large domestic supplies and a robust coal infrastructure would continue to explain its commitment to coal as it transitioned to renewables. Large domestic supplies of coal may make it more difficult for Germany to significantly reduce its dependence on fossil fuels.

To counteract the insecurities caused by the volatility in oil prices in the 1970s, Germany began to transition to a nuclear power portfolio. With the nuclear meltdown at the Fukushima power facility in Japan, national policy expanded its commitment to wind and solar power. That coal-rich Germany, without domestic supplies of oil and natural gas, would opt for renewables suggests its commitment to reducing its carbon footprint. An environmental tradition in Germany with a vocal minority green political party helps explain the transition to wind and solar rather than an expansion of nuclear power with its potential for cataclysmic events and the storage for its spent radioactive fuel. Germany decided to reduce and eventually phase out its nuclear program in the early 2000s. Sluggish demand, low prices and a positive energy security outlook favored the phase-out. Support for renewable energy flourished given its support from a political coalition and a continued dependence on domestic coal for electrical generation.

The diffusion of technology played a significant role in the adoption of wind power in Germany. It adopted wind energy technology from Denmark, a global leader in wind power, and grafted it onto its own energy research and development schemes. This diffusion of knowledge across borders aided the transition to renewables. While Germany supported wind turbine research in the 1970s and 1980s, it failed to produce a viable design and abandoned the project. In 1990, a coalition of political parties including the Greens and the Social Democratic Party passed Feed-in-tariffs (FITs) requiring utilities to purchase electricity from small hydropower plants at 90 percent of the retail price. Unexpectedly, the law resulted in a 100-fold increase in wind power in the 1990s, representing 1 percent of Germany's electricity generation by 1999.[32]

This initial foothold in wind power encouraged some German manufacturers to enter the wind turbine market. In the early years of the twenty-first century, the wind power industry became the second largest in the world, behind the United States. Individuals and cooperatives began investing in wind energy installations. Again, the coalition of Greens and SDP strengthened the FITs for 20 years for wind and solar PV. In 2001, wind turbines produced 3.5 percent of Germany's electricity. As the world's third-largest economy with 80 million people, Germany planned to supply 28 percent of its energy from onshore and offshore wind by 2030.

Asian wind power

China

As the world's largest consumer of energy in 2018, its demand for energy will double by 2030. With the world's second-largest reserves in coal, some of which is inaccessible in remote regions of the country and too expensive to extract and transport, China became a net importer of coal in the first half of 2007. Its imports of coal, oil and natural gas will continue to rise in the future. In terms of China's national security, dependence on energy imports poses risks to its future economic development. Given China's large size and its long coastline, the potential energy from wind may alleviate the country's dependence on imports of fossil fuels. According to China's Meteorological Administration, onshore wind resource potential is 1,000 gigawatt-hours, and the offshore resource is 300 gigawatt-hours. While growth has been slow in recent years, China began to develop wind power in the early 1970s. Its primary purpose then was to provide electricity in remote pastoral areas inhabited by herdsmen and women and isolated islands with no connection to an electrical grid.

Not until the late 1980s did grid-connected wind power begin to add electricity to China's growing demand. The country bought small wind turbines from Vestas, a Danish supplier capable of producing between 55 and 150 kilowatt-hours of electricity. With generous financial support from provincial and the central government, wind farms began to appear throughout the country using imported wind turbines. Throughout the 1990s and into the next century, government policy insured wind power's commercial success by providing financial security for investors and subsidizing both the construction of wind farms and the establishment of favorable electrical rates for consumers.

By the end of 2000, however, installed capacity missed the national target of 1 gigawatt-hours set by the former Ministry of Electric Power. At that date, installed capacity had reached 344 megawatt-hours only. Formerly, government policy focused on purchasing wind turbines from foreign suppliers and incentivizing investors to build wind farms. In the new century, government policy promoted the domestic manufacturing of wind turbines. By 2005, wind power supply reached 1,260 megawatt-hours. While still missing the goal of 1 gigawatt-hours, China became the world's fastest-growing wind market with an annual growth rate of 56 percent.[33]

Replacing coal as a major source of energy remains a daunting undertaking by the Chinese government. Yet, its investments in renewable energy grew by 100 times compared to 2005. Its goal of boosting wind power installation to 250 gigawatt-hours and solar to 150 gigawatt-hours by 2020 seems achievable. Currently, China accounts for one-third of the world's investment in renewable energy. In recent years, however, the acceleration of renewable energy installations faced two unpredictable obstacles. China's rapid economic growth stalled, and the failure to modernize and extend the transmission grid system curtailed the increasing use of renewables. "From 2010 to 2016, 150.4 million megawatt-hours, or as much as 16 percent of overall wind generation, was abandoned. The total energy loss is equivalent to 48 million tons of coal consumption, or 134 million tons of carbon dioxide emissions."[34] Despite these losses, China became the leading producer of wind power in 2010, replacing the United States. China's installed capacity has doubled every year since 2006 while its total exploitable capacity for onshore and offshore wind is between 700 and 1,200 gigawatt-hours.

As the projected annual need for electricity grows by 10 percent, China will require an additional 800 gigawatt-hours of electricity for the next 20 years. Reverting to coal-fired generation would add 3.5 gigatons of carbon dioxide emissions each year through 2030. Substituting electrical power from wind at a cost of US$0.059 per kilowatt-hour could displace 23 percent of coal-generated electricity. This change could eliminate as much as 0.62 gigatons of annual carbon dioxide emissions.[35]

India

India's coastline stretches 7517 km from the Bay of Bengal to the Arabian Sea in the Indian Ocean. Its territorial waters extend up to 12 nautical miles into the ocean. These characteristics make India a potential wind powerhouse. With the air over the Indian landmass heating more rapidly during the day, it expands and rises. Rising cooler ocean air rushes in to fill the space, creating winds. At night, the process is reversed, with the air over the land cooling more rapidly than that over the Indian Ocean, creating more wind.

Figure 9.3 Ancient Royal Cenotaphs (Residence) at Jaisalmer, India, with modern windmills.

India's geographic location and its extensive coastline create the conditions for a potential wind energy electrical generation of 65,000 megawatt-hours. With that potential, it helps to explain India's incentive to build a substantial wind energy industry. The Indian industry is 30 years old and now ranks as the fourth-largest wind market globally, with an installed capacity of more than 31 gigawatt-hours. Having reached that goal in March 2017, India's target of reaching 60 gigawatt-hours by 2022 seems achievable. The government has made substantial commitments to meet the terms of the Paris Agreement and follow a cleaner path than those at a similar level of economic development. To achieve this goal, private investment accounts for almost 90 percent of the financial support for wind energy.

With a population exceeding 1.2 billion, 240 million, or 20 percent of the population, remain without electricity. Cutting in half the number of people without electricity and a doubling of electricity for rural populations places the 20 percent in perspective. It suggests momentum toward bringing renewable electricity to more citizens. In 2016, 3.6 gigawatt-hours of installed capacity decreased the size of the population without electricity, and in March 2017, 5.4 gigawatt-hours of new installations decreased the numbers further.

The momentum to increase India's share of renewable energy and achieve the goal of raising its share of non-fossil-fuel power capacity to 40 percent by 2030 will require both clarity in its commitment and a stable regulatory framework. The country's Integrated Energy Policy projects 800 gigawatt-hours of installed capacity by 2031–2032, with 320 gigawatt-hours, or 40 percent, coming from renewable energy. Meeting its 2020 goal requires the installation of 60 gigawatt-hours of wind energy and 100 gigawatt-hours of solar capacity. The same mix may help India meet the Paris Agreement of 40 percent renewables by 2030.

Promising developments in recent years point toward an acceleration in the deployment of renewable energy resources. The current manufacturing capacity of wind turbines in India is about 10 gigawatt-hours, with foreign companies building blade-manufacturing factories in states with high wind-energy potential. The Indian state of a Gujarat, one of many such states, has two such factories built by LM Wind Power. Vestas, another wind turbine manufacturer, opened its factory there in 2017. With an estimated wind potential of 84.4 gigawatt-hours, domestic and foreign wind turbine manufacturers will continue to invest in Gujarat's wind potential.

Offshore wind potential

The commitment made in the Paris Agreement substitutes renewable non-fossil-fuel energy from coal, oil and natural gas to 40 percent from solar and wind by 2030. A renewed commitment to offshore wind energy suggests a rapid expansion of electrification. By 2050, renewable energy resources could account for more than 80 percent of India's energy needs. According to Alok Kumar, India, country manager for energy, "[o]ffshore wind could play a very important role in India. With a coastline of over 7,600 km, India offers tremendous potential for offshore wind."[36] Offshore wind faces fewer constraints than land-based turbines. With most of the onshore sites already exploited, finding new sites becomes more difficult. In addition, coastal cities with large populations will become the beneficiaries of offshore wind power. Its higher and steadier winds make meeting the 2050 targets possible. The states of Gujarat and Tamil Nadu possess eight geographical zones most suitable for offshore wind development. With access to current technology, both states have the potential of each producing 100 gigawatt-hours, with offshore wind farms much larger than onshore installations. By 2019, the Offshore Wind Policy Board will begin accepting bids for offshore platforms in the two key coastal states of Gujarat and Tamil Nadu.

Tamil Nadu

This southern Indian state's wind turbines generate 7.9 gigawatt-hours of electricity, putting it ahead of countries committed to green energy. Many of these countries have smaller populations than this Indian state, such as Sweden, which aims to generate its energy from renewable sources. Its current wind capacity is 6.7 gigawatt-hours. The Institute for Energy Economics and Financial Analysis suggests that Tamil Nadu could double its wind power capacity to 15 gigawatt-hours by 2027. Not to be minimized, Indian states added more wind power than

Tamil Nadu. They included Andhra Pradesh (2.2 gigawatt-hours) and Gujarat (1.3 gigawatt-hours). Combined, all Indian states generate more renewable energy from wind than any other source, including solar.[37]

Barriers to wind power development

Despite the enthusiastic support from India's central government, it shares political power with state governments, some of whom don't support central government initiatives with enthusiasm. Both exercise control with states and local utilities oftentimes playing a larger decision-making role than the government in Delhi. State power exists despite their questionable financial status and their inability to make timely payments to state utilities. To compensate for delays in payment, the central government plans to make loans to distressed state utilities for performance-related matters. Also, a problem for India in making the transition from fossil fuel energy sources to renewables is its ability to transfer power from windy states to other parts of the country. Creating an effective interstate electricity market requires a reliable infrastructure with transfer stations, poles and transmission lines. "Proper incentives would be a major boost to the renewables sector and the Indian power system as a whole."[38] As with most countries, there's a long road ahead in the transition from fossil fuels to renewables. India currently uses coal to generate more than 60 percent of its electricity.

Conclusion

The future of wind turbines is reflected in its powerful growth. By the end of 2020, the total world wind turbine capacity will reach 817 gigawatt-hours an increase of almost 68 percent over 2016 levels. This accelerated growth makes it the fastest-growing alternative energy resource in the world. Wind turbines deliver increasing amounts of energy in countries, including the United States, Denmark, Germany, China and India. This sample of countries globally represents a fraction of those invested heavily in renewable energy and those planning to begin the transition by seeking financial support from developed countries.

The Power Africa Initiative with US$7 billion coming from the United States in research and development funds over five years will help build the West Africa Power Pool. With technical support from Gigawatt Global, an Israeli firm, the installation of 800 megawatt-hours of wind and solar power will begin in the West African countries of Burkina Faso, Senegal, Mali, Nigeria and Gambia. Creating a regional energy electricity grid will begin the process of addressing energy poverty in less developed countries.

Technical roadblocks, environmental and aesthetic considerations remain obstacles as the industry for wind turbines matures. Initially, technical difficulties plagued the industry. Wind turbines experienced excessive maintenance, limiting their hours of operation and the amount of power produced. Both liabilities limited the ability of wind power to compete with the cost of fossil fuels. Over time, technical support eliminated many of these deficiencies. The challenge remains, however, in extending the life of wind turbines and reducing operating and maintenance costs. With multi-megawatt turbines having about 8,000 individual components, many linked to the drive train, wear and tear in gearboxes and bearings will require predictive maintenance. Recognizing the unique vibration signals of a turbine's many moving parts will alert skilled technicians of the need to remain vigilant.

Wind turbines emit no greenhouses gases, nor do they produce radioactive waste. Comparatively, their minimal environmental impact makes them a viable alternative to fossil fuels and nuclear power. Their real and potential damage to birds remains a hazard for flocks traveling along genetically defined flyways. Available solutions exist to limit bird deaths that wind turbines formerly killed in the thousands. Avian Radar in Texas warns operators to stop turbines when it signals the presence of birds. Studying the breeding and feeding behavior of birds informs planners of areas to avoid in building wind turbines. For offshore wind farms, protecting fish and marine mammals will require similar research initiatives.

Mechanical and aerodynamic noise have been identified with sleep disorders and hearing loss among residents living close to wind farms. Also, some environmental engineers suggest that the turbulence caused by wind turbines may cause local climate change by mixing the air up and down. The turbulence may also change the direction of high-speed wind, causing the evaporation of moisture on the ground.[39]

With wind becoming a reliable and sustainable source of power to generate electricity, research regarding its immediate and long-term environmental impact remains an area for further inquiry. Two studies suggest the need for more research. In Inner Mongolia, data about rainfall noted that an unprecedented drought since 2005 occurred faster in areas with large wind farms. In the Coachella Valley of Southern California, the San Gorgonio wind turbines could change local temperatures by warming surface temperatures at night and cooling them during the day.[40] Future studies about the environmental impact of wind turbines will hopefully reveal more about their effects.

Notes

1 www.atlasobscura.com/places/nashtifan-windmills.
2 www.telosnet.com/wind/early.html.
3 Martin Pasqualetti, Robert Righter, and Paul Gipe, "Wind Energy, History of," in *Encyclopedia of Energy*, eds. Cutler J. Cleveland and Robert U. Ayers, vol. 6 (Amsterdam: Elsevier Inc., 2004), 420.
4 Wilson Clark, *Energy for Survival: The Alternative to Extinction* (Garden City, New York: Anchor Books, 1974), 521.
5 Pasqualetti, et al., "Wind Energy, History of," 421.
6 Ibid.
7 Ad van der Woude, "Sources of Energy in the Dutch Golden Age," in *Economia e Energia, SECC. XIII-XVIII, Instituto Internazionale Di Storia Economica*, ed. Simonetta Cavaciocchi (Le Monnier, 2002), 446.
8 Ibid., 447.

9 Adam Lucas, *Wind, Water, Work: Ancient and Medieval Milling Technology* (Leiden: Brill, 2006), 126.
10 Ibid.
11 Rene Leboutte, "Intensive Energy Use in Early Modern Manufacture," in *Economia e Energia*, 559–560.
12 Vaclav Smil, *Energies* (Cambridge, MA: MIT Press, 1999), 125.
13 www.telosnet.com/wind/early.html.
14 Pasqualetti, et al., "Wind Energy, History of," 421.
15 Joel A. Tarr, "Lighting the Streets, Alleys and Parks of the Smoky City: Networked and Non-Networked Technologies, 1878–1928," Unpublished manuscript, 2018.
16 "Mr. Brush's Windmill Dynamo," *Scientific American* 63, no. 25 (December 20, 1890), 389.
17 Pasqualetti, et al., "Wind Energy, History of," 422.
18 Paul Gipe, *Wind Power: Renewable Energy for Home, Farm, and Business*, 3rd ed. (White River Junction, VT: Chelsea Green Publishing Company, 2004), 13.
19 "Science-Harnessing the Wind," *Time*, September 8, 1941.
20 Pasqualetti, et al., "Wind Energy, History of," 424.
21 Matthias Heymann, "Signs of Hubris: The Shaping of Wind Technology Styles in Germany, Denmark, and the United States, 1940–1990," *Technology and Culture* 39, no. 4 (October 1998), 13.
22 Pasqualetti, et al., "Wind Energy, History of," 425.
23 Dennis Y.C. Leung and Yuan Yang, "Wind Energy Development and its Environmental Impact: A Review," *Renewable and Sustainable Energy Reviews* (November 4, 2011), 1034.
24 Frank Jossi, "Industry Report: Midwest and Great Plains Lead Wind Energy Expansion," https://energynews.us/2017/04/19/midwest/industry-report-midwest-and-greatplains-lead-wind-energy-expansion/.
25 Ibid., 431.
26 Ibid., 425.
27 Ibid., 5–6.
28 Frode Birk Nielsen, "A Formula for Success in Denmark," in *Wind Power in View: Energy Landscapes in a Crowded World*, eds. Martin Pasqualetti, Paul Gipe and Robert Righter (New York: Academic Press, 2002), 117.
29 Ibid., 117–123.
30 "Denmark Sets World Record for Wind Power," *Climate Action in Partnership with UN* (January 19, 2016).
31 Aleh Cherp, Vadim Vinichenko, Jessica Jewell, et al., "Comparing Electricity Transitions: A Historical Analysis of Nuclear, Wind and Solar Power in Germany and Japan," *Energy Policy* (November 24, 2016): 616–617.
32 Ibid., 620.
33 Xia Changliang and Song Zhanfeng, "Wind Energy in China: Current Scenario and Future Perspectives," *Renewable and Sustainable Energy Reviews* 13 (2009): 1968–1970.
34 Ye Qi, Jiaqi Lu and Mengye Zhu, "Wind Curtailment in China and Lessons from the United States," *Brookings* (March 1, 2018), 1.
35 Michael B. McElroy, Xi Lu, Chris P. Nielsen, and Yuxuan Wang, "Potential for Wind-Generated Electricity in China," *Science* 325 (September 11, 2009): 1379.
36 Sapna Gopal, "Offshore Winds Could Bring Change to India's Renewable Energy Sector," *Mongabay India* (April 27, 2018): 4.
37 John McKenna, "This Indian State Produces More Wind Power than Sweden and Denmark," *World Economic Forum* (February 21, 2018): 2–5.
38 Dwivedi, Rishi, Lauha Fried, Steve Sawyer and Shruti Shukla, "Indian Wind Energy-A Brief Outlook," *Global Wind Energy Council* (2016): 10.
39 Leung and Yang, "Wind Energy Development and Its Environmental Impact," 1037.
40 Ibid.

References

Changliang, Xia, and Song Zhanfeng. "Wind Energy in China: Current Scenario and Future Perspectives." *Renewable and Sustainable Energy Reviews* 13, no. 8 (October 2009): 1966–1974. doi: 10.1016/j.rser.2009.01.004.

Cherp, Aleh, Vadim Vinichenko, Jessica Jewell, et al. "Comparing Electricity Transitions: A Historical Analysis of Nuclear, Wind and Solar Power in Germany and Japan." *Energy Policy* 101 (February 2017): 612–628. doi: 10.1016/j.enpol.2016.10.044.

Clark, Wilson. *Energy for Survival: The Alternative to Extinction*. Garden City, New York: Anchor Books, 1974.

"Denmark Sets World Record for Wind Power." *Climate Action in Partnership with UN*, January 19, 2016.

Dwivedi, Rishi, Lauha Fried, Steve Sawyer and Shruti Shukla. "Indian Wind Energy-A Brief Outlook." *Global Wind Energy Council* (2016): 10.

Gipe, Paul. *Wind Power: Renewable Energy for Home, Farm, and Business*. White River Junction, VT: Chelsea Green Publishing Company, 2004.

Gopal, Sapna. "Offshore Winds Could Bring Change to India's Renewable Energy Sector." *Mongabay India*, April 27, 2018, 4.

Heymann, Matthias. "Signs of Hubris: The Shaping of Wind Technology Styles in Germany, Denmark, and the United States, 1940–1990." *Technology and Culture* 39, no. 4 (October 1998): 641–670.

Jossi, Frank. "Industry Report: Midwest and Great Plains Lead Wind Energy Expansion." Accessed November 5, 2018. https://energynews.us/2017/04/19/midwest/industry-report-midwest-and-greatplains-lead-wind-energy-expansion/.

Leboutte, Rene. "Intensive Energy Use in Early Modern Manufacture." In *Economia e Energia, SECC.XIII-XVIII, Instituto Internazionale Di Storia Economica*, edited by Simonetta Cavaciocchi, 547–575. Firenze, 2002.

Leung, Dennis Y.C., and Yuan Yang. "Wind Energy Development and its Environmental Impact: A Review." *Renewable and Sustainable Energy Reviews* 16, no. 1 (November 4, 2011): 1031–1039. doi: 10.1016/j.rser.2011.09.024.

Lucas, Adam. *Wind, Water, Work: Ancient and Medieval Milling Technology*. Leiden: Brill 2006.

McElroy, Michael B., Xi Lu, Chris P. Nielsen, and Yuxuan Wang. "Potential for Wind-Generated Electricity in China." *Science* 325 (September 11, 2009): 1378–1380. doi: 10.1126/science.1175706.

McKenna, John. "This Indian State Produces More Wind Power Than Sweden and Denmark." *World Economic Forum*, February 21, 2018, 2–5.

"Mr. Brush's Windmill Dynamo." *Scientific American* 63, no. 25 (December 20, 1890): 38.

Nielsen, Frode Birk. "A Formula for Success in Denmark." In *Wind Power in View: Energy Landscapes in a Crowded World*, edited by Martin Pasqualetti, Paul Gipe and Robert Righter, 115–132. New York: Academic Press, 2002.

Pasqualetti, Martin, Robert Righter, and Paul Gipe. "Wind Energy, History of." In *Encyclopedia of Energy*, vol. 6, edited by Cutler J. Cleveland and Robert U. Ayers. Amsterdam: Elsevier Inc., 2004.

Qi, Ye, Jiaqi Lu and Mengye Zhu. "Wind Curtailment in China and Lessons from the United States." *Brookings*, March 1, 2018, 1.

"Science-Harnessing the Wind." *Time*, September 8, 1941.

Smil, Vaclav. *Energies*. Cambridge, MA: MIT Press, 1999.

Tarr, Joel A. "Lighting the Streets, Alleys and Parks of the Smoky City: Networked and Non-Networked Technologies, 1878–1928." Unpublished manuscript, 2018.

van der Woude, Ad. "Sources of Energy in the Dutch Golden Age." In *Economia e Energia, SECC.XIII-XVIII, Instituto Internazionale Di Storia Economica*, edited by Simonetta Cavaciocchi, 444–468. Firenze, 2002.

Part IV

Alternative energy solutions

10 Fuel cells and battery power

Reducing greenhouse gas emissions

Introduction

Two possible energy sources to curtail greenhouse gas emissions are hydrogen fuel cells and storage batteries for vehicles and the electrical grid. Using gasoline-to-power vehicles emits 20 pounds of carbon dioxide for each gallon of gasoline burned.[1] It is the most energy inefficient use of fossil fuel.

> For every twenty gallons put into a car we only get two gallons of actual work. The rest is dissipated as heat. The other inherent inefficiency is in the fact to move one person, we have to spend energy moving over a ton of material (the automobile itself).[2]

Fuel cells convert the chemical energy found in hydrogen into electrical energy in one step. Conversely, the internal combustion engine requires several steps. First, the engine must convert the chemical energy found in gasoline into thermal energy by combustion. Thermal energy expands the gases found in gasoline within the vehicle's cylinders. The high-pressure gas is then converted to mechanical energy by the pistons and the drive train. With so many energy conversions, much of the power of internal combustion engines is lost to incomplete combustion and exhaust heat. With an efficiency rate of 20 percent or less for most internal combustion engines, their replacement would be welcome by both climate scientists and the driving public.[3] The number miles driven by consumers globally increased by 1.3 percent in 2017, or 32 billion miles or 170 round trips from Earth to the Sun![4] Reducing emissions for automobiles and trucks could curb climate change.

Improvements in storing the energy from sunlight and wind for the stationary electrical grid using battery power will reduce emissions as well. Electrical Energy Storage (EES) converts electrical energy into a stored form that can later be converted when needed into electricity. Globally in 2018, installed energy storage reached 175.8 gigawatt-hours. In Europe, 10 percent of its energy was provided by energy storage facilities, and in Japan, it reached 15 percent. EES technologies described here focus on electrical grids. They include storage technology types, including Pumped Hydroelectric Storage (PHES) and Compressed Air Energy

Storage (CAES). Several other battery storage projects include those based on lead-acid, lithium-ion, nickel and sodium-based and flow batteries.[5]

Hydrogen fuel cells

As a replacement for gasoline vehicles, scientists and engineers project that fuel cell vehicles will become two to three times more efficient. Relative to other ways to generate electricity, fuel cells have the advantages of low costs with high efficiencies. Simply stated, a fuel cell takes oxygen from the air and hydrogen from a tank, creating a chemical reaction to produce water vapor and electric energy.[6] Typically, fuel cells burn hydrogen that reacts with oxygen in the air to produce electricity and water. Neither produces heat-trapping gases that warm the global climate system. As the most abundant element found on Earth's surface, hydrogen is locked in water and plants. Despite its abundance, hydrogen is always combined with other elements, such as water as in H_2O.

By passing an electrical current through water, it can split water into its components, H_2 and O, and compress and store the hydrogen. When the Sun disappears below the horizon and the winds stop blowing, the stored hydrogen becomes a power source.[7] This separation process is known as electrolysis. A single fuel cell makes electrons using an electrolyte located between two electrodes. Plates on the side of the cell collect electrical current and discharge gases. Their versatility permits stacking from a few to many fuel cells, depending on their application. Conventional batteries store electrons and with every recharge lose power. Since hydrogen fuel cells make electrons, power is not lost.[8]

Unlike batteries, however, fuel cells will continue to produce electricity and never lose their charge as long as the fuel source is hydrogen. "It can be produced from water through renewable forms of energy – including sun, wind, water, geothermal energy, and biomass – that are readily available in every country."[9] Since hydrogen-powered fuel cells do not pollute, they offer one potential solution to curbing environmental pollution. Hydrogen-powered fuel cell vehicles with a battery and an electric motor could double automotive fuel efficiency. Producing hydrogen through low-carbon-emitting processes and establishing a functional distribution system, comparable to the network of existing gas stations, would make possible low-greenhouse-emissions hydrogen fuel cell vehicles. Achieving such outcomes will require technological innovations and years before fuel cell vehicles become commonplace.

Early history of the hydrogen fuel cell

The history of this technology began when a Welsh judge, chemist and physicist, Sir William R. Grove, invented a gas voltaic battery to produce electricity from hydrogen and oxygen. He reported his resulted in the *Philosophical Magazine* in 1839. He half immersed the hydrogen and oxygen electrodes in a bath of diluted sulfuric acid to produce electricity. This was achieved by decomposing the water by electrolysis. Sir William wrote, "This is to my mind, the most interesting effect

Figure 10.1 This hydrogen fuel cell generates enough electricity to power an automobile with a driving range of 550 kilometers (342 mi), eliminating the need for gasoline.

of the battery; it exhibits such a beautiful instance of the correlation of natural forces."[10] In this experiment, Sir William used hydrogen as fuel and pure oxygen, or air, as oxidant.

The absence of a need for electricity delayed further experimentation. Fifty years later, however, the chemists Ludwig Mond and Charles Langer, in the United Kingdom, searched for a cheaper way to produce electricity. Since plentiful coal produced impure coal gas, Mond and Langer substituted it for either air or pure oxygen. Although the results in producing electricity proved to be promising, using coal gas, rather than hydrogen, emitted carbon dioxide poisoning the process. Mond's research and attention to refining his coal gas experiment got sidetracked when he discovered that extinguishing the coal-gas flame with a cold porcelain tile produced a shiny metallic nickel mirror. The birth of this new nickel industry resulted in a change in Mond's attention, pushing fuel cell research aside. As founder of the International Nickel Company, his research had increased the power density of the fuel cell.

Thomas Edison's coal-burning generating station built in Lower Manhattan in 1882 converted about 2.5 percent of the energy available into electricity. Into the twentieth century, steam turbines received just fewer than 20 percent. As early as 1894, Wilhelm Ostwald pointed out the wastefulness of the steam engine and provided motivation for fuel cell research. He envisioned an emissions-free modern society by eliminating burning fossil fuels with no smoke and no soot.[11] Providing electricity and heat for households, factories and commercial buildings and electric motors for automobiles, trucks and buses would become a reality in the not-too-distant future.

Perceived efficiencies became commonplace as stoves became ways to direct energy into households and factories, with wood, coal, coal gas, oil and natural gas providing illumination and heat. Interest in fuel cell technology waned, even though research in Switzerland in the early years of the twentieth century kept it alive. Sir Francis T. Bacon, an engineering professor at Cambridge University, noted that the commercial success of fuel cells would have to operate at ordinary temperatures using hydrogen as a fuel source. In 1932, Bacon developed the first practical 5 kilowatt-hour hydrogen–oxygen alkaline electrolyte system. In a modified state, it powered an Allis Chalmers agricultural tractor, and in partnership with the Army Air Force, it powered a welding machine, a circular saw and a 2-ton forklift.[12]

Space age applications of fuel-cell technology

Engineers solved the problem of the corrosion of fuel cells by adding lithium for improved electronic conductivity. When the Soviet Union launched its small satellite, Sputnik, in 1957 into orbit around Earth, the "space race" commenced. This event and those that followed greatly influenced the development of fuel cells. Once the National Aeronautics and Space Administration (NASA) selected the fuel cell over other competing power systems, research and development experienced almost explosive growth at universities and industry. For extended missions with humans on board, the power, efficiency, weight and safety of fuel cells outweighed all competing systems. The U.S. Gemini, Apollo and Space Shuttle programs used this technology for powering its spaceships. As an added benefit, the hydrogen–oxygen fuel cell supplied potable water for the crew's consumption, a by-product of electrochemical reaction and air cabin humidity.[13]

Beginning in the late 1950s and early 1960s, NASA, in partnership with engineers Willard Thomas Grubb and Leonard Niedrach at General Motors, invented the proton exchange membrane fuel cell (PEMFC). It provided only modest cell voltage and low current for the 200 hours required to support *Gemini*'s four short orbital missions. Further experiments led to the development of alkaline fuel cells with higher cell voltage and much-improved energy used in the Apollo space missions. The space shuttle's fuel cell development program represented a vast improvement over the Apollo program in every spacecraft application. All systems used pure hydrogen–oxygen because of its promise to provide safer, more flexible and efficient power requirements for NASA's lengthy missions. Choosing the hydrogen fuel cell over all others caused an explosive growth in research and development sponsored by the government, carried out in its laboratories and in industries and universities.

NASA's space shuttle fuel-cell technology represented a vast improvement over the Apollo program. Higher cell voltage and a much lower weight improved energy and power greatly. Space missions originally intended to last 2,500 hours, but with vehicles built to accomplish multiple missions, NASA needed to find ways to extend the energy and power capabilities of its future Shuttle missions. Before shifting to a new power array based on the proton exchange membrane (PEM), minor modifications of the shuttle's existing hydrogen–oxygen fuel cells doubled its life to 5,000 hours. To get beyond this 5,000-hour boundary, however, PEM fuel cells would be needed to provide the 10,000 hours of energy and power for the solar-powered Skylab and the International Space Station.[14] Much of NASA's research and development of fuel cell technology focused on spacecraft applications with an emphasis on its human space flight program with few commercial applications.

Green power and the coming hydrogen economy

A series of distressing events triggered a heightened awareness of the air pollution caused by coal-fired plants, factories and urban sprawl that increased vehicular traffic. Two developments in the 1970s brought public awareness of the need to consider alternative sources of energy for motorized vehicles. Further dependence on fuel to support an increasingly motorized population became evident with the oil embargoes of 1973 and 1979 by nations in the Organization of Petroleum Exporting Countries. Gasoline shortages, runaway fuel prices and a polluted atmosphere became drivers of fuel cell research and development. National governments and automobile manufacturers began research projects to develop efficient fuel cell technologies.

An emerging environmental movement in the 1970s became the backdrop for exploiting the potential of a hydrogen-based transportation system. As an alternative to fossil gasoline, the research initiative in hydrogen–oxygen fuel cells focused on the polluting characteristics of the internal combustion engine. Kurt Weil, a retired professor from Stevens Institute of Technology, put it succinctly when he stated, "It is not the internal-combustion engine that pollutes our air, but its present fuels . . . the hydrogen-burning internal-combustion engine already exists in practical and proven models."[15] The systemic impact of air pollution on public health, air quality and infrastructure led many researchers to express their outrage at the environmental effects of internal combustion.

One report cast the importance of this transition from one energy source to another in dire, almost cataclysmic terms:

> Future historians may well record that the private automobile and not the nuclear bomb was the most disastrous invention of a society so obsessed with technology that it never recognized the failures of engineering run rampant until too late . . . survival demands that the air pollution caused by the automobile be eliminated almost immediately.[16]

During the 1970s, university researchers and those in private companies used hydrogen either alone or in combination with gasoline to reduce unburned hydrocarbons emissions. An economic recession in the 1980s stalled commercial applications. As an example, President Ronald Reagan (1981–1989) removed the solar panels from the roof of the White House installed by President James Earl Carter (1977–1981). This event symbolized the government's denial of the atmospheric pollution caused by burning fossil fuels.

By the mid-1990s, however, the political environment began to change. "Green Power" and "The Hydrogen Economy" led to commercial investments in fuel cell technology that outstripped investments made previously by NASA. As advances in space technology spearheaded NASA, automobile propulsion systems would become a research focus of government and the automotive industry. A South African electrochemist, John O'Mara Bockris, first used the term *hydrogen economy*, believing that hydrogen would be used as an energy source to replace or substitute for a large use of fossil fuels.

NASA's goals differed from those of automobile manufacturers. Human exploration of the solar system required the development of a unique high-energy system with the highest reliability needed for a mission's successful accomplishments. The importance of developing a unique system for human space flight and developing fuel cells for an automobile market for hundreds of thousands of units (automobiles) did not fit well with NASA's goals and that of the automobile industry. However, "the Hydrogen Economy" with its long-term investment in "green energy" requires a NASA/Department of Energy/industry partnership. Cooperation remained the key to sustaining both efforts.[17]

Hydrogen fuel cell transportation

As noted earlier, extremely expensive materials, including platinum limited a wider application of fuel cells, all of which required very pure hydrogen and oxygen. High operating temperatures turned out to be a limiting factor as well. Not until the 1990s did several innovations drive down costs. Reduced use of platinum and thinner film electrodes resulted in wider applications focused on transportation. Other applications included stationary and portable power generation. The lightweight of PEMFCs made them suitable for buses that used compressed hydrogen for fuel and operated at 40 percent efficiency. By the mid-2000s in Europe, China and Australia high efficiency, zero-emissions fuel cell buses became attractive for vehicles running on established routes that could be regularly fueled with hydrogen at their home garages. Costs affected the use of fuel cell buses. They are five times more expensive to build than diesel buses, and the start-up cost of building a hydrogen infrastructure may be prohibitive. If cities decide, however, that efforts to reduce emissions that cause urban pollution are more costly to public health and their built environment, the transition fuel cell vehicles make economic and social sense.

"Hydrogen will be the most important energy source of the 21st century. Long term, it will replace oil and gas." So claimed Fritz Vahrenholt, an executive at Deutsche Shell, the German division of Royal Dutch/Shell Group. His response was an acknowledgment of the opening of Europe's first commercial hydrogen refueling station on January 13, 1999, in Hamburg, Germany. According to the city's mayor, Ortwin Runde,

> [t]he streets will be quiet. Only the sound of tires and rushing wind will accompany passing vehicles instead of the roar from exhaust pipes. The city will be clean, since emissions will be practically zero. Pedestrians strolling on the sidewalks won't be turning up their noses, guests won't be fleeing from the street's stench into the cafes because now they can enjoy their sundowners in the open air.[18]

Although such enthusiasm may be unwarranted with the opening of one hydrogen refueling station, its intent was to gain support for a new emission-free transportation system. The opening covered by Germany's television stations and CNN reinforced the mayor's message. The cost of hydrogen fuel cell power about 13 cents higher per kilometer (0.62137 mi) than gasoline and about 4 to 5 cents more than diesel would be offset by buying commercially produced hydrogen made from renewable sources such as geothermal and hydroelectric power imported from Iceland.[19]

With European, Japanese, Korean, and U.S. automobile manufacturers entering the marketplace of advanced transportation technologies in the twenty-first century, the earlier skepticism about hydrogen-fueled buses, trucks and automobiles evident earlier seems to have weakened. New Flyer Industries in Winnipeg, Manitoba Province, Canada, built a 60-passenger low-floor bus with a range of 250 miles using hydrogen technology. The next year, 1995, the Chicago Transit Authority tested three hydrogen-powered PEM-fuel-cell buses. The city's environment department commissioner, Henry Henderson, argued that hydrogen technology could alter urban transportation into the twenty-first century. Test sites in Vancouver and California's Coachella Valley used up-to-date fuel cell technologies.

In Europe, fuel cell buses debuted in 1994 with the launching of the Eureka, an 80-passenger vehicle. Air Products of Netherland produced the buses liquid-hydrogen fuel system. Thirty Daimler Chrysler fuel cell buses provide service to passengers in 10 European cities with plans to introduce them in Beijing, China and Perth, Australia. In London, Millennium Taxi, introduced its first combined 5-kilowatt fuel cell and batteries in experimental black cabs. In the same year, Germany's NEBUS (New Electric Bus) made technological advances. It contained a 10-stack 250-kilowatt PEM fuel cell, enough for traction, electrical systems and air-conditioning. To suggest that its complicated design could be easily replicated would be an understatement. On its roof, 300 gas bottles held 45,000 liters of compressed hydrogen, enough to power the bus for 156 miles. A simpler commercial version became available in 2000 with a smaller fuel-cell system placed on the roof to provide better equilibrium, weight distribution and more room for passengers.[20]

At the same time, Daimler Chrysler fleet of 60 hydrogen fuel-cell cars, the F-Cell became available for global testing. In different driving conditions, testing this real-world car to perform, as a nonpolluting and energy-efficient vehicle became a priority. These conventional looking cars have competitors, as 20 of Honda's FCX and 30 of Ford's Focus FCV fuel-cell cars appear on the horizon. Name a leading auto manufacturer – Toyota, Nissan, Renault, Volkswagen, Mitsubishi and Hyundai – and you have new entrants into the competition for the most efficient fuel-cell vehicles.

Currently, manufacturers have built thousands of fuel cell vehicles for trials globally. With governments imposing stricter regulations on exhaust emissions that cause climate change, the automobile industry and governments have invested several tens billions of dollars in developing a clean and efficient propulsion system for vehicles that will replace the internal combustion (IC) engine. Skeptics reply that a consumer-ready fuel cell vehicle remains a wish without a comprehensive long-term plan. Proponents see no alternative to these experimental vehicles. Neither advances in IC technology nor hybrid vehicles provide an alternative solution. Hybrid vehicles combine IC technology with electrochemical batteries. They burn gasoline that produces carbon dioxide and other hydrocarbons that cause atmospheric pollution.

Yet, several stumbling blocks will delay the marketing fuel cell–powered vehicles. Currently, none of the prototypes have the fuel storage capacity that achieves the 300- to 400-mile range of IC cars to which consumers have become accustomed. If they did approach this minimum range, the current infrastructure for fueling stations and technical maintenance would limit vehicle utility. The entire network of gas stations would need to be overhauled. A General Motors study estimated that it would cost US$10 billion to US$15 billion to build 11,700 new fueling stations in the United States, or one every 2 miles in metropolitan areas and one every 25 miles on interstate highways.[21] Figuring out how to increase onboard fuel capacity, lower the cost of their drive trains to a hundredth of current costs, increase the life of vehicles fivefold and power up the energy output for SUVs and trucks remains problematic. "So despite bright hopes and the upbeat pronouncements of automakers, considerable technical and marketing challenges remain that could delay introduction of the fuel-cell family car for years, if not decades."[22]

Scientists believe that the transition to a "hydrogen economy" will take decades to accomplish because of the problems noted earlier and those that follow. Key to this transition is the production process. Making hydrogen from methane, a greenhouse gas, violates the goal of reducing and eventually eliminating atmospheric warming. Also, if electrolysis, the process by which splitting water into hydrogen and oxygen uses fossil energy, then it produces carbon dioxide, a greenhouse gas. Finally, using fossil fuel to produce hydrogen spends more energy than it produces in the resulting hydrogen. How to produce more hydrogen using less energy remains a significant challenge.

High-temperature electrolysis, running electricity through water heated to 1000 degrees Celsius (1832 °F) breaks up the water molecules. A ceramic sieve separates the oxygen from the hydrogen. Although heating the water requires more energy than that gained by the separation process, it's more efficient than all other known processes. Removing impurities from industrially produced hydrogen limits its use for vehicles. Will all the preceding be solved? Science, technology and public investment solved problems associated with building a national railroad system in the nineteenth century and an interstate highway system at the mid-twentieth century. Given this historical perspective, optimism in solving the vehicular fuel cell remains intact.[23]

Conclusion

Until the price of vehicles powered by fuel cells declines, it is unlikely that they will become vehicles for the masses. Toyota's fuel cell Mirai retails for US$57,500, a price beyond the budgets of most citizens.[24] To cut costs, researchers from laboratories and businesses are experimenting with less expensive catalysts. Currently, commercial catalysts consist of thin layers of platinum nanoparticles placed on a carbon film. As noted earlier, the high price and scarcity of precious metal platinum prevent the widespread production and sale of fuel cell vehicles. Finding a cheaper and more plentiful catalyst such as palladium and shrinking or replacing platinum remains an intermediate goal. Using a cheaper and more plentiful metal, such as copper or nickel, as a catalyst may be a feasible long-term solution. Ultimately, scientists believe that a catalyst without metals would allow fuel cells to compete with the next-generation batteries.[25]

A battery-powered future

Energy and power provided determine the value of batteries. Energy defines the amount of work that can be completed and power tells us how quickly that work gets done. In addition to hydrogen–oxygen fuel cell technology, research and development on long-lasting batteries offer another opportunity to break our fossil fuel dependence. Batteries are getting better, but like fuel cells, progress is slow. Rechargeable lithium-ion batteries dominate the world of smartphones, but their instability and threat of explosion limit their utility for larger tasks.[26] Providing electricity to power vehicles, households and factories remains a promised but as-yet-unrealized goal. The focus on wind and solar energy without an efficient energy storage component remains a plan without a benefit. How to capture and retain energy produced by wind and solar when the sun disappears below the horizon and the wind stops blowing becomes a conundrum without storage capacity.

Traditionally, our lithium-ion batteries did not deliver the storage needs of the future. In use today to power our computers and cars, lithium-ion batteries have become slimmer and smarter. Upscaling their storage capacity has led to electric cars that drive longer on a single charge. Their liabilities include high cost, weight, explosiveness when you try to power them up and the equally great shortcoming of ease of proper disposal. While computer processors double their capacity every two years, in the past, lithium batteries eked out a few percentage points of improvement in a similar two-year horizon. Improvements and efficiencies could come much faster. Possibly, a solid-state lithium-ion battery with a solid rather than liquid electrolyte would be more powerful. Many are in use as automobiles become a potential market. Despite the limitations of lithium-ion batteries, their life in powering electric cars has reached 310 miles on a single charge for Tesla's Model S 100D.[27]

Lithium

Lithium is the 31st most abundant chemical element found on Earth, usually in granite deposits that contain 1.0 to 2.5 percent concentrate. In Australia and China, granite ore must be mined, crushed into sizable chunks, liquefied, washed and dried to produce powdered lithium. Using this process, the Greenbushes Mines in Australia store and ship lithium to an expanding global market. In the Andes Mountains of Peru, minerals leeching into the soil over millions of years produced sulfate, potassium, boron and lithium. Exposing the minerals by pumping water into saline brine ponds produces liquid lithium deposits. Exposure to sunlight evaporates the saltwater ponds, leaving only the liquid lithium deposits. The Australian, Chinese and Peruvian methods represent a few of the processes used for releasing lithium for global markets. This plentiful element will be available for a lengthy period to help power small electronics, vehicles, and possibly electric grids. If a lithium crunch does occur, it will be after the production of hundreds of millions of Tesla-class electric vehicles.[28]

Conventional batteries possess the anode connected to a cathode by an electrolyte but over time become depleted. Lithium is highly conductive but unstable. As an electrically charged compound, it is safer. Lithium-ion batteries are usually made of laminated structures with a material at their center called an electrolyte. The ions move back and forth between negative and positive electrodes when the batteries are charging or discharging. Graphite is commonly used for the negative electrode coupled with a metal oxide in the positive electrode in a lithium-ion battery. By replacing the graphite with lithium metal, the metal electrode could increase a battery's cell-specific energy by 35 percent and its energy density by 50 percent. Such a battery cell could deliver 350 to 400 kilowatt-hours per kilogram and 1,000 kilowatt-hours per liter. Current battery packs with graphite have a specific energy of 150 kilowatt-hours per kilogram and an energy density of 250 kilowatt-hours per liter. Using solid materials might one day make the lithium-ion battery obsolete. A solid electrolyte would make the battery less explosive and safer. Such a technological breakthrough would revolutionize vehicular performance and a grid storage capacity much greater.[29]

A rechargeable molten-metal battery

"A battery will do for the electricity supply chain what refrigeration did to our food supply chain," according to Donald Sadoway, professor of material science at Massachusetts Institute of Technology (MIT). His goal is to develop a new low-cost, high-capacity, long-lasting, rechargeable molten-metal battery. He and his collaborators want to make it successful for storing energy on power grids

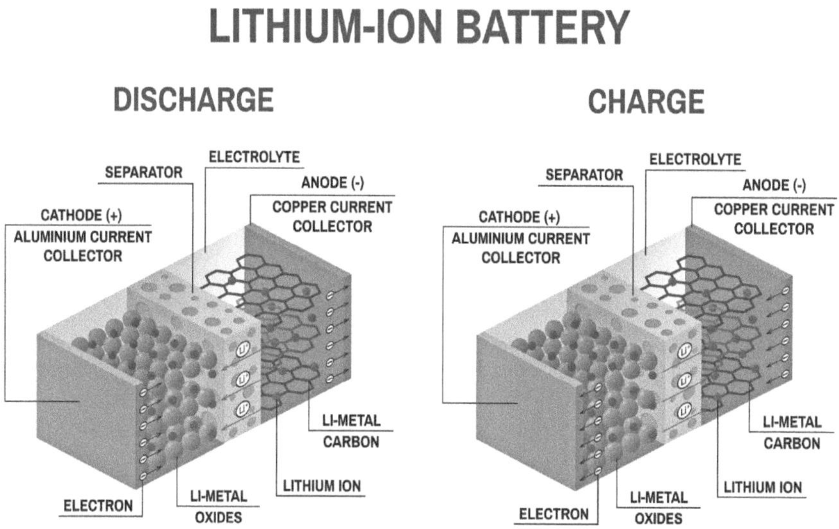

Figure 10.2 Diagram of a lithium-ion battery.

now and in the future. Using Earth-abundant materials, his high-capacity batteries using molten metals eliminates the problems of lithium-ion batteries losing capacity and degrading with repeated charging using solid electrodes. Traditionally, lithium-ion batteries are small and upgraded every few years to avoid obsolescence. The electrical grid, on the other hand, requires large, reliable batteries that operate for years without interruption.

Such batteries would address the age-old problem of storing large amounts of energy from solar and wind to be delivered when needed. Large storage capacity would curtail the climate change caused by burning fossil fuels. As Professor Sadoway noted, "[i]n the energy sector, you're competing among hydrocarbons, and they are deeply entrenched and heavily subsidized and tenacious."[30] Facing a competitive disadvantage, the cost of production, and an issue of concern for hydrogen–oxygen fuel cells as well, he addressed this issue first rather than following the classic academic model of creating the best product imaginable and making costs secondary. A liquid metal battery became the solution to the problem of grid-scale storage.

Liquid metal lithium batteries

Similar to conventional batteries, Sadoway's liquid-metal lithium-ion battery contains electrodes at the top and bottom with an electrolyte between. All three are liquid. The top layer, a negative electrode, is a low-density liquid metal. It contributes electrons to the battery's positive electrode bottom layer, a high-density liquid metal that readily accepts the electrons. The electrolyte middle layer is molten salt that transfers charged particles but does not mix with either layer because of their differences in density. When the battery is discharging, the top layer of molten metal gets thinner and the bottom layer gets thicker. Negative electrons from the top layer travel through the electrolyte to the high-density bottom layer. Once there, they pass through an external circuit, powering the electrical load that moves the vehicle or provides grid-scale storage. Recharging the battery turns the thickness processes in reverse, beginning with the bottom layer moving through the middle electrolyte layer to the low-density top layer.

All these exchanges from discharging to recharging happen ultrafast because the components are liquid. With the quick flow of large electrical currents into and out of the lithium-ion battery, liquid metal remains pliable, reducing stress. Over time, solid electrodes in conventional lead-acid batteries degrade and often crack, resulting in mechanical failure.[31] In 2010, Sadoway founded a battery company, Ambri, with former students to manufacture molten-metal battery packs. After successfully dealing with problems with the heat seals, the packs reach self-sustaining operating temperatures: hot enough to charge and discharge without extra energy input.[32]

Flow-battery technology combines the characteristics of both fuel cells and batteries where liquid energy creates electricity and is recharged simultaneously. It is a rechargeable battery that uses the spent electrolyte liquid and recovers it for reenergization. "The fundamental difference between conventional batteries and flow cell batteries is that the energy is stored as the electrode material in conventional batteries but as electrolyte in flow cells."[33]

Sodium-ion batteries

The lithium-ion molten metal battery is one of many developments in the search for increased storage capacity. Another battery developed by MIT researchers is called sodium-ion, a battery to extend its life longer than its companion the lithium-ion. Unlike the lithium-ion battery where charging and discharging causes electrode materials to shrink and expand at a rapid rate that may eventually cause it to crack, sodium-ion batteries can be fine-tuned to change the range of the expansion and contraction of the electrodes, thereby increasing their life. Since lithium is a mined element, shortages will increase its cost per ton. Salt, however, is much more plentiful and literally cost-free. The sodium-ion battery suggests that other materials previously considered unusable might offer engineers opportunities to design new generations of batteries.

As noted earlier, U.S. consumers burn more than 9 million barrels of gasoline each day. The energy consulting firm Wood Mackenzie projects that if electric cars gain more than 35 percent market share by 2035, the United States could see a cut from 9 million to 2 million barrels a day. These plug-in electric vehicles (PEVs) would offer zero tailpipe emissions when operating solely on battery power. Compared to gasoline-powered vehicles, the high cost of electric cars remains the major barrier to adoption. Developing a battery that would cost

US$150 per kilowatt-hour of energy storage remains the goal. A kilowatt-hour of energy output sells for about 10 cents and will move a car 3 to 4 miles. Currently, lithium-ion batteries cost US$1,000 per kilowatt-hour of energy output. Bringing the cost down to US$150 is the theoretical low as projected by engineers.[34]

Assumptions about the future costs of electric, gasoline and hybrid vehicles follow, knowing that the volatility in the costs of electricity and the costs of gasoline can't be known. Given these assumptions, the fuel for an electric vehicle with an energy efficiency of 3 miles per kilowatt-hour costs about 3.3 cents per mile when electricity costs 10 cents per kilowatt-hour. The fuel for a gasoline vehicle with an energy efficiency of 22 miles per gallon costs about 15.9 cents per mile when gasoline costs US$3.50 per gallon. The mileage for commercial fleet vehicles such as light-duty pickups ranges from less than 17 miles per gallon to generally about 22 miles per gallon. The energy cost per mile is also included for a hybrid electric vehicle (HEV) with an energy efficiency of 45 miles per gallon, as these types of vehicles are increasingly being used. If US$3.50 per gallon of gasoline were also assumed for the HEV that gets 45 miles per gallon, the energy cost per mile would be 7.8 cents per mile.[35]

With the price of lithium-ion batteries dropping by 90 percent between 1990 and 2015, a continued downward trend seems inevitable. The cost of electric vehicle batteries continues a downward spiral, lower than the costs projected in the last decade. By 2020, it is conceivable that Tesla's "gigafactory" will manufacture US$100/kilowatt-hour lithium-ion batteries for electric vehicles. The growth of the electric vehicle (EV) market will continue to accelerate with the declining prices of batteries. Reducing transportation's carbon footprint will get us closer to carbon-free energy. Lithium-ion fuel cells remain the best current option for powering electric vehicles. As stable and compact batteries, however, they are still too bulky and expensive despite improvements during the last two decades. The storage capacity of a 6-inch-square pack has tripled from 200 per hour to more than 700 watt-hours. To power an electric automobile, a fuel cell battery must generate 50 to 100 kilowatt-hours of electricity, so the tripling of a battery pack's power exceeds the daily needs of most drivers. Its weight and storage capacity are also compatible with the size of current electric vehicles. They must have at least one-half ton in a space equal to the size of a conventional automobile trunk. So, advances in battery power meet many of the requirements that we associate with the transition from conventional internal combustion engines to those powered by electricity. The cost of that electricity at about US$150 per kilowatt-hour, however, exceeds the U.S. Department of Energy's standard of US$100 for affordability. As noted, innovations in battery power suggest a trend in which costs will continue their downward spiral. Meeting cost requirements that make affordability the norm will not be achievable without attention to the cost and availability of scarce materials.[36]

Those scarce materials, nickel and cobalt, used in the electrodes of lithium-ion batteries exceed the demand in a high-growth market with surging battery production. Before demand depletes global supplies of cobalt and nickel, research and development need to find cheap, accessible alternative substitute metals for electrodes. Common metals such as copper or iron fluorides and silicon store lithium ions by bonding chemically with them. Early-stage research and development look promising yet stability, charging speeds and production remain problems to be solved. Without a viable solution, the transition from internal combustion engines that produce seventy percent of the world's air pollution to electric cars will be curtailed.[37] As discussion of the economic viability of energy storage for the grid develops, the benefits of both solar and wind technology will become clearer.

Storage batteries for the electrical grid

In the early decades of this century, all the world's functional large-scale lithium-ion batteries combined stored less than a minute of the world's demand for electricity. As the size and number of these batteries increase, prices for a kilowatt-hour of electricity will invariably drop. Today, storing less than a minute of electricity costs about 25 cents per kilowatt-hour. Once storage capacity reaches 13 minutes of electricity demand, lithium-ion prices will have dropped to the 10 to 13 cents per kilowatt-hour stored. Once storage capacity reaches an hour of demand for electricity, cost will have dropped to the range of 6 to 9 cents per kilowatt-hour. Once a full day of storage is achieved, prices will be down to 2 to 4 cents and become competitive with natural gas prices that do not factor in the cost of carbon dioxide emitted. More important, batteries used by utilities allow them to reduce the cost of new or refurbished transmission and distribution lines. In addition, utilities could maximize their lines during low peak hours to recharge batteries close to customers. Recharged batteries reduce the need for transmission and distribution lines during peak hours. For example, in California, peak-hour prices compared to nighttime prices range from 34 cents to 14 cents. This significant difference suggests the viability of storage batteries.[38]

Research, development and deployment of state-of-the-art lithium-ion batteries continue unabated. One such deployment provides a case study of storage solving energy shortages for South Australia, one of the country's five states. A series of extreme weather events including more than 80,000 lightning strikes and two tornadoes crippled the state's electrical grid dependent on Australia's abundant reserves of coal. An exceptionally hot and dry summer placed additional stress on an overworked power grid, causing blackouts and brownouts. To lessen the burden placed on the state's 1.7 million residents, the government made energy storage part of a AUS$550 million energy plan. A contract to produce batteries to store 100 megawatts became an integral part of future deployment. Tesla's battery division won the contract in competitive bidding, claiming that it would produce batteries in 100 days that would store 100 to 300 megawatts.[39]

For Elon Musk, Tesla's chief executive officer who built his reputation by developing and selling the Tesla Model S and X electric vehicles, developing and deploying batteries for the grid represented a new challenge. Yet, not surprisingly, Musk's reputation for delivering in the electric vehicle business, suggested that he would not have proposed and won a contract if he believed it was unachievable in 100 days. By November 23, 2016, 40 days before the deadline, his Nevada factory delivered the power packs that created a 100-megawatt-hour/129-megawatt-hour battery bank that can store enough energy to power 30,000 homes in South Australia for an hour. Although it represented the world's largest lithium-ion battery installation, its number one status would be eclipsed soon thereafter in South Australia by a much larger installation of 1.1 million batteries storing the energy collected by 3.4 million solar panels.[40]

Compressed Air Energy Storage (CAES)

CAES facilities compress ambient air and store it in an underground container to respond to sudden demands for electricity. Conventional coal plants cannot respond to demands quickly and the lack of hills precludes the construction of hydroelectric plants. In Huntorf, Germany, a 321-megawatt-hour utility built a compressed air storage facility. It owned land that contained ancient underground salt beds that it dissolved by pumping water into them. The process produced two caverns about a half-mile below the utility's grassy fields. Salt caverns are preferable to other belowground caverns because they do not lose pressure within the storage container and no negative reaction occurs with oxygen in the air or in the salt rock. Operational since 1978, the plant uses power from Germany's grid when demand is low and cheaper. It uses this power to compress and store air into the caverns. When demand for power surges, a motor pushes the compressed air (oxygen) to the surface and into a natural gas combustion chamber (a furnace) where it spins a turbine producing electricity. Compressed air contains more oxygen as it enters the turbines, making the production of electricity more efficient.[41]

A 110-megawatt-hour facility built in McIntosh, Alabama, uses the same CAES technology as the one in Huntorf, Germany. New plants in the planning stage use similar technology with an innovation. Newer-generation plants are mostly conventional gas turbines that separate the compression of the combustion air from the natural gas turbine process. As such, the newer energy storage plants have two additional benefits. By uncoupling the compression stage from the turbine stage, the CAES can produce three times the power output using the same amount of natural gas. Reducing natural gas output lowers carbon dioxide emissions by upward of 40 to 60 percent if the waste heat is used to warm up the compressed air. A more efficient adiabatic method recaptures the heat of compression and reheats it during turbine operations, thereby reducing the volume of natural gas needed to warm the decompressed air. By doing so, it is 70 percent more efficient than other methods and reduces carbon dioxide emissions further.[42]

Thermal energy storage

Thermal energy storage is another general category that represents advanced technology. Modern solar thermal power plants produce all their energy when the sun shines. Excess energy produced during peak sunlight hours is stored in molten salt cavities to be used during off-peak nighttime hours with low energy demands. Released to generate steam that drives a turbine, it produces electricity. With lower electricity rates at night, the stored sunshine can be used to produce ice that can become an important part of a building's cooling system during the daytime. Thermal energy storage works much like a well-designed thermos that stores the energy in a coffee mug or a cooler.

Pumped Hydroelectric Storage (PHES)

PHES is connected to two large steel thermal tanks filled with gravel-grade crushed rock as a storage medium. A closed circuit filled with gas such as argon connects the two tanks, one that serves as a compressor and the other as an expander. Stored electrical energy drives a heat pump. It pumps argon into the top of the compressed air thermal tank. The argon is compressed to +500 degrees Celsius (+932 °F) and flows slowly through the gravel (for storage), heating it as the gas cools. As the gas exits at the bottom of the compressed thermal tank and enters the bottom of cold expander tank, it cools to minus −160 degrees Celsius (−256 °F). The argon flows slowly upward in the expander tank cooling the gravel as the gas warms, leaving the top of the tank at ambient temperature.

To recover the energy that's been discharged, the entire heating to cooling process is reversed. At room temperature, the argon enters the expander thermal tank and flows downward, heating the gravel (storage medium) as the gas cools. It exits the tank at −160 degrees Celsius and enters the compressor. There, the gas is heated to +500 degrees Celsius. Following the steps described earlier, once the hot, pressurized gas then enters the expander thermal tank, it releases the energy compressed in the gas to drive the heat pump. Using gravel as a storage medium provides a very low-cost storage solution.[43]

Hydrogen energy storage

Electrolysis can convert electricity in hydrogen, which has a much higher storage capacity than small batteries or large scale CAES. It can be re-electrified in fuel cells with high efficiencies and stored in pressurized tanks and liquefied at −253 degrees Celsius (−423 °F). Underground salt caves can store large amounts of hydrogen at very high densities. With such a large number and diverse set of storage facilities available, the production of energy to meet daily and seasonal variations can be accommodated efficiently.

As we have seen, the development of electrolytic hydrogen for future fuel cell cars is an ongoing project. Other uses include the mixing of hydrogen with natural gas of up to 15 percent in natural gas pipelines and as a feedstock for the chemical and petrochemical industries. Hydrogen can also be used in the production of synthetic liquid fuels from biomass, adding further to the utility of biomass as a fuel source. The industry standard is to store large volumes of hydrogen in salt caverns. Currently, two exist in the United States, in Texas, with a third one under construction. Three others operate in Teesside, United Kingdom.[44]

Pumped hydropower

Pumped hydropower is the most common form of energy storage using dams, a large-scale gravity storage technology. The hydroelectric dam holds moving river water and releases it down a spillway and through a turbine to create electricity for a major or local grid. Pumps may also be used to elevate water into a retaining pool behind a dam. In this way, water can be released on demand to activate turbines below the dam to produce electricity.

Although stored water has been the primary energy resource in producing hydropower, changes in the global climate system have resulted in some decades-long droughts in some areas of the world. The use of gravel as a substitute for water serves the needs of public and private electric power–producing utilities. The gravitational system works by moving gravel up the side of a hill to be stored and then released on demand. The weight of the gravel activates a mechanical system the drives a turbine to produce electricity. To date, gravitational systems have the largest energy storage capacity based on produced megawatts.[45]

Subsurface pumped hydroelectric storage

Without dammed rivers, lakes and ocean water, storage to be released on demand to produce electricity could not exist. However, the unpredictability of a changing climate that amplifies the vagaries of weather has suggested ways to avoid both depleted and overflowing reservoirs. Although relatively new, the concept of underground, subsurface, reservoirs reflects efforts to reduce the potential hazards posed by climate change. The availability of existing sites including mines, natural caverns and the potential to build subsurface reservoirs by excavation and construction creates available and potential sites.

The cost of creating new subsurface sites may be economically prohibitive unless the excavated material has financial value. Mines and caverns remain the focus of developers. Three such planned projects use existing sites, two of which are described here. The first located in Elmhurst, Illinois, about 20 miles from Chicago and called the Elmhurst Quarry Pumped Storage (EQPS) Project would operate in the following way. Surface water will be gravity fed to an underground facility. During peak hours, water would be released to a pump turbine that would generate electricity and deliver it to the power grid aboveground. Temporarily, the water would be held in a mine or cavern and pumped to the surface during low-cost off-peak hours. When completed, the energy storage potential is 708.5 gigawatt-hours.[46]

Riverbank Wiscasset Energy Center (RWEC) proposes a 1,000-megawatt-hour pumped storage facility located 2,200 feet underground in Wiscasset, Maine. Ocean water would be dropped into an underground shaft down 2000 vertical feet into a powerhouse. There, four 250-megawatt-hour pump turbines would be ready to receive the water to generate electricity for the grid during peak hours. As with EQPS, the water would be stored in underground reservoirs ready to be pumped to the surface using off-peak power.[47]

Conclusion

As recently as 2018, energy storage capacity represented a fraction of the total global energy output. Its growth rate in recent years suggests that the innovative designs of new and more powerful batteries will become commonplace globally. Efforts to capture and store energy from the sun and the wind may begin to cut into the ever-declining role of coal and the current dominant roles of oil and natural gas in powering the global economy. In 2018, the United States deployed 41.8 megawatts of storage capacity. Again, only a fraction of the country's energy needs but a 46 percent growth over the previous year, with each quarter showing an average 10 percent change.[48]

Pumped hydro storage capacity accounted for 95 percent of global installed capacity until recently.[49] Compressed air storage, a more recent technological innovation, has begun with three commercial plants in Germany, the United Kingdom and the United States, with a total capacity of 600 megawatt-hours. More plants will come online as countries move from development and construction to installation. Increasing battery storage capacity is reflected in both small and large installations. In Germany, around 25,000 domestic installations with a total capacity of 160 megawatt-hours will grow to 250,000 by 2020. Millions of domestic batteries store the sun's energy to operate water heaters. They provide a substantial benefit by shifting 5 percent, or about 20 terawatt-hours, from peak to low-demand periods. As more domestic capacity becomes available, the energy savings and environmental benefits will mount.[50]

Worldwide, large commercial installed batteries produce more than 750 megawatt-hours of electricity, with the number increasing as lithium-ion, sodium and flow technologies become more popular. Flow batteries may begin to dominate the field. From 2015 to the present, global installed storage capacity consisting of 944 projects generating 146 gigawatt-hours has increased to more than 1050 projects producing 274 gigawatt-hours of electricity. The United States leads the field in installed capacity with 227 lithium-ion installations producing 473 megawatt-hours, 135 thermal producing 664 megawatt-hours and 38 pumped hydropower (dams) producing 22,561 megawatt-hours of electricity.

Thermal energy storage may come to dominate the field for large-capacity commercial projects. China, Japan and South Korea, in that order, follow in storage capacity, ranging from 54 to 44 lithium-ion installations. With substantial growth in electric vehicles, China's capacity will continue to grow as it becomes the world's largest economy. Four hydrogen energy storage plants producing 3 megawatt-hours of electricity supplement Germany's 35 lithium-ion projects. China generates more than 23,000 megawatt-hours of pumped hydro storage, with the United States following closely behind with 22,500 megawatt-hours.[51]

Large battery systems serve to eliminate the variability in power supply caused by wind and solar power. Can batteries installed at a data center smooth out the demand and decrease in demand that occurs each day from wind and sun energy sources? A solution would eliminate the need for utilities to build additional natural gas plants to match the increasing demand for power. By powering data centers with wind and solar turns them into data plants, something between natural gas power plants and data centers. Unlike conventional natural gas power plants that often take time to meet spikes in demand, data plants with energy-storage batteries dispatch power immediately. Since technology companies buy and use lots of electricity, they represent the vanguard to developing data plants. Google claims that its global power consumption equals that of the city of San Francisco. Instead of buying power from the grid, these companies power their own enterprises and sell excess power to utilities. Microsoft's Boydton facility in Virginia is equipped to store photographs, videos and web-based news programs when it loses power from the grid. Can technology companies serve its backup needs and simultaneously provide power to the grid? With such high stakes for buyers of electricity, a solution awaits more planning and execution. One corporate energy strategist noted, "These companies are showing they have an in-depth understanding of energy markets, but they want to help decarbonizes the grid while maintaining resiliency, security and all those other factors."[52]

Currently, the lithium-ion battery dominates the vehicular and electrical generating grid but are there other competitors that challenge its primary space? A recent 2017 poll of 500 professionals at Greentech's energy storage summit answered the following question: What technology has the best chance of supplanting lithium-ion as the dominant utility-scale advanced storage technology? Of the participants, 46 percent identified flow batteries while only 23 percent identified lithium-ion. Twenty-two percent indicated that some other, yet-to-be-developed energy storage system would supplant all others.[53]

Notes

1 "How Much Gasoline Does the United States Consume?" The U.S. Energy Administration, www.eia.gov/tools/faqs/faq.php?id=23&t=10.
2 "Energy System," http://environ.andrew.cmu.edu/m3/s3/all_ene_sys.htm.
3 Noriko Behling, "Making Fuel Cells Work," *Issues in Science and Technology* (Spring, 2013), 83.
4 U.S. Dept. of Transportation, Federal Highway Administration; "Traffic Volume Trends," (September 2017), 1. 17septvt.pdf.
5 "U.S. Grid Energy Storage Factsheet," *Center for Sustainable Systems, University of Michigan* (August 2018).
6 Jeff Tollefson, "Fuel of the Future," *Nature* 464 (April 29, 2010): 1262.
7 ReNew, "Fuel Cells: The Key to Cleaner Transport," (January-March 1998), 39.
8 www.driveclean.ca.gov/Search_and_Explore/Technologies_and_Fuel_Types/Hydrogen.
9 Behling, "Making Fuel Cells Work," 84.
10 F.T. Bacon and T.M. Fry, "The Development and Practical Application of Fuel Cells," *Proceedings of the Royal Society of London. Series A, Mathematical and Physical Sciences* 334, no. 1599 (September 25, 1973): 430.
11 M.L. Perry and T.F. Fuller, "A Historical Perspective of Fuel Cell Technology in the 20th Century," *Journal of the Electrochemical Society* 149, no. 7 (2002): S59.
12 Marvin Warshay and Paul R. Prokopius, "The Fuel Cell in Space: Yesterday and Tomorrow," *Journal of Power Sources* 29 (1990): 193.
13 Warshay and Prokopius, "The Fuel Cell in Space," 193.

14 John H. Scott, "Fuel Cell Development for NASA's Human Exploration Program, Benchmarking with 'The Hydrogen Economy,'" *Journal of Fuel Cell Science and Technology* (July 15, 2007): 1–3.
15 Peter Hoffmann and Tom Harkin, *Tomorrow's Energy: Hydrogen, Fuel Cells, and the Prospects for a Cleaner Planet*, 123.
16 Ibid., 135.
17 Ibid., 6–7.
18 Hoffmann and Harkin, *Tomorrow's Energy*, 99.
19 Lain Staffel, Daniel Scamman, Anthony Velazquez Abad, et al., "The Role of Hydrogen and Fuel Cells in the Global Energy System," *Energy & Environmental Science* (December 10, 2018).
20 Ibid., 121.
21 Tollefson, "Fuel of the Future," 1263.
22 Steven Ashley, "On the road to Fuel Cell Cars," *Scientific American* (2005), 63–64.
23 Ibid., 68–69.
24 www.carscoops.com/2016/09/2017-toyota-mirai-fuel-cell-retails-for/.
25 Donna J. Nelson, "Hyrogen Cars for the Masses," *Scientific American* 317, no. 6 (December 2017): 36.
26 M. Bini, D. Capsoni, S. Ferrari, et al. "Rechargeable Lithium Batteries: Key Scientific and Technological Challenges," *Woodward Publishing Series in Energy* (2015): 1–17.
27 Liqiang Mai, Mengyu Yan, and Yunlong Zhao, "Track Batteries Degrading in Real Time," *Nature* 546, no. 7659 (June 21, 2017): 469.
28 Ramez Naam, "How Cheap Can Energy Storage Get? Pretty Darn Cheap," at http://rameznaam.com/2015/10/14/how-cheap-can-energy-storage-get/.
29 Paul Albertus, Susan Babinec, Scott Litzelman, and Aron Newman, "Status and Challenges in Enabling the Lithium Metal Electrode for High-Energy and Low-Cost Rechargeable Batteries," *Nature Energy* 3 (January 2018): 16–17.
30 Nancy Stauffer, "A Battery Made of Molten Metal: New Battery May Offer Low-Cost, Long-Lasting Storage for The Grid," *Energy Futures* (December 14, 2015), http://energy.mit.edu/energy-futures/autumn-2015/.
31 Ibid.
32 Amelia Urry, "Inside the Race to Build the Battery of Tomorrow," *Wired* (February 22, 2017): 6.
33 *Energy Storage Association*, "Flow Cells," at http://energystorage.org/energy-storage/storage-technology-comparisons/flow-batteries.
34 For comparison, the average household in the USA, the average use of power is about 750 kilowatt-hours per month or 9,000 kilowatt-hours per year. This is 9 megawatt-hours per year. A megawatt-hour is a measure of total energy used over a period. One megawatt is equivalent to a million watts of power used for one hour. It is the measure used in buying and selling power between power providers and utilities. Household bills are not measured in megawatts.
35 "Comparing Energy Costs per Mile for Electric and Gasoline-Fueled Vehicles," Advanced Vehicle Testing Activity, Idaho National Laboratory, https://avt.inl.gov/sites/default/files/pdf/fsev/costs.pdf.
36 Katie Fehrenbacher, "The Promise and Challenge of Scaling Lithium Metal Batteries," *Greentech Media: A Wood Mackenzie Business*, www.greentechmedia.com/articles/read/lithium-metal-battery-promise-challenge?
37 Ibid., 470.
38 Ramez Naam, "The Price of Energy Storage is Plummeting," at http://rameznaam.com/2015/04/14/energy-storage-about-to-get-big-and-cheap/#Grid.
39 Hi Zeng Feng Sun, G.E. Weichun, et al., "Introduction of Australian 100MW Storage Operation and Its Enlightenment to China," *China International Conference on Electricity Distribution* (September 17–19, 2018): 2896–2897.

40 David Roberts, "Elon Musk bet that Tesla Could Build the World's Biggest Battery in 100 Days. He Won." *VOX* (December 20, 2017): 1–6.
41 Diane Cardwell, "The Biggest, Strangest 'Batteries," *The New York Times* (June 4, 2017): 6.
42 *Energy Storage Association*, "Diabatic CAES Method," http://energystorage.org/compressed-air-energy-storage-caes.
43 http://energystorage.org/energy-storage/technologies/pumped-heat-electrical-storage-phes.
44 http://energystorage.org/energy-storage/technologies/hydrogen-energy-storage.
45 http://energystorage.org/energy-storage-techologies/pumped-hydro-power.
46 http://energystorage.org/energy-storage-techologies/sub-surface-pumped-hydroelectric-storage.
47 Ibid.
48 "U.S. Grid Energy Storage Factsheet."
49 Shafiqur Rehman, Luai M. Al-Hadhrami, and M.D. Mahbub Alam, "Pumped Hydro Energy Storage Systems: A Technological Review," *Renewable and Sustainable Energy Review* 44 (2015): 586–598.
50 www.degruyter.com/view/j/green.2011.1.issue-3/green.2011.027/green.2011.027.xml.
51 http://rameznaan.com/2015/10/14/how-cheap-can-energy-storage-get/.
52 Benjamin Storrow, "How Big Batteries at Data Centers Could Replace Power Plants," *E&E News* (July 19, 2018): 4.
53 Dan Finn-Foley, "The Next 5 Years in Energy Storage, According to 500 Energy Professionals," *Greentech Media* (January 11, 2018): 3.

References

Advanced Vehicle Testing Activity, Idaho National Laboratory. "Comparing Energy Costs per Mile for Electric and Gasoline-Fueled Vehicles." Accessed November 5, 2018. https://avt.inl.gov/sites/default/files/pdf/fsev/costs.pdf.

Albertus, Paul, Susan Babinec, Scott Litzelman, and Aron Newman. "Status and Challenges in Enabling the Lithium Metal Electrode for High-Energy and Low-Cost Rechargeable Batteries." *Nature Energy* 3 (January 2018): 16–17.

Ashley, Steven. "On the Road to Fuel Cell Cars." *Scientific American*, March 2005, 63–64.

Bacon, F.T., and T.M. Fry. "The Development and Practical Application of Fuel Cells." *Proceedings of the Royal Society of London. Series A, Mathematical and Physical Sciences* 334, no. 1599 (September 25, 1973): 427–452. doi: 10.1098/rspa.1973.0101.

Behling, Noriko. "Making Fuel Cells Work." *Issues in Science and Technology* 29, no. 3 (Spring, 2013).

Cardwell, Diane. "The Biggest, Strangest 'Batteries.'" *The New York Times*, June 4, 2017, 6.

Energy Storage Association. "Diabatic CAES Method." Accessed November 5, 2018. http://energystorage.org/compressed-.

Energy Storage Association. "Flow Cells." Accessed November 5, 2018. http://energystorage.org/energy-storage/storage-technology-comparisons/flow-batteries.

Finn-Foley, Dan. "The Next 5 Years in Energy Storage, According to 500 Energy Professionals." *Greentech Media*, January 11, 2018. www.greentechmedia.com/articles/read/the-next-five-years-in-energy-storage-according-to-500-energy-professionals#gs.xdOSbIE.

Hoffmann, Peter, and Tom Harkin. *Tomorrow's Energy: Hydrogen, Fuel Cells, and the Prospects for a Cleaner Planet*. Cambridge, MA: MIT Press, 2014.

Mai, Liqiang, Mengyu Yan, and Yunlong Zhao. "Track Batteries Degrading in Real Time." *Nature* 546 (June 21, 2017): 469–470. doi: 10.1038/546469a.

Naam, Ramez. "How Cheap Can Energy Storage Get? Pretty Darn Cheap." Accessed November 5, 2018. http://rameznaam.com/2015/10/14/how-cheap-can-energy-storage-get/.

Naam, Ramez. "The Price of Energy Storage is Plummeting." Accessed November 5, 2018. http://rameznaam.com/2015/04/14/energy-storage-about-to-get-big-and-cheap/#Grid.

Nelson, Donna J. "Hyrogen Cars for The Masses." *Scientific American* 317, no. 6 (December 2017): 36.

Perry, M.L., and T.F. Fuller. "A Historical Perspective of Fuel Cell Technology in the 20th Century." *Journal of the Electrochemical Society* 149, no. 7 (2002): S59–S67. doi: 10.1149/1.1488651.

Roberts, David. "Elon Musk Bet That Tesla Could Build the World's Biggest Battery in 100 Days. He Won." *VOX*, December 20, 2017, 1–6.

Scott, John H. "Fuel Cell Development for NASA's Human Exploration Program, Benchmarking with 'The Hydrogen Economy." *Journal of Fuel Cell Science and Technology* (July 15, 2007): 1–8.

Stauffer, Nancy. "A Battery Made of Molten Metal: New Battery May Offer Low-Cost, Long-Lasting Storage for the Grid." *Energy Futures* (Autumn, 2015). http://energy.mit.edu/energy-futures/autumn-2015/

Storrow, Benjamin. "How Big Batteries at Data Centers Could Replace Power Plants." *E&E News*, July 19, 2018, 4.

Turcheniuk, Kostiantyn, Dmitry Bondarev, Vinod Singhal, et. al. "Ten Years to Redesign Lithium-Ion Batteries." *Nature* 559 (July 25, 2018): 467–470. doi: 10.1038/d41586-018-05752-3.

Urry, Amelia. "Inside the Race to Build the Battery of Tomorrow." *Wired*, February 22, 2017, 6.

Warshay, Marvin, and Paul R. Prokopius. "The Fuel Cell in Space: Yesterday and Tomorrow." *Journal of Power Source* 29 (1990). doi: 10.1016/0378-7753(90)80019-A.

Index

Note: Page numbers in *italics* refers to figures.

FSC
www.fsc.org
MIX
Papier | Fördert
gute Waldnutzung
FSC® C083411

Kolibri 360 GmbH
Ferdinand-Jühlke-Straße 7
99095 Erfurt, Deutschland
produktsicherheit@kolibri360.de